Technology and the Rest of Culture

Technology and the Rest of Culture

Edited by Arien Mack

The Ohio State University Press
Columbus

Originally published in 1997 as a special issue of *Social Research*.
First Ohio State University Press edition published 2001.

Library of Congress Cataloging-in-Publication Data

Technology and the rest of culture / edited by Arien Mack.
　　p. cm.
　　Includes bibliographical references.
　　ISBN 0-8142-0869-X (alk. paper)—ISBN 0-8142-5070-X (pbk. : alk. paper)
　　1. Technology—Social aspects. I. Mack, Arien.

T14.5 .T44187 2001
303.48′3—dc21

00-047890

Jacket and cover design by David Drummond.
Printed by

The paper used in this publication meets the minimum requirements of
the American National Standard for Information Sciences—
Permanence of Paper for Printed Library Materials.
ANSI Z39.48-1992.

9 8 7 6 5 4 3 2 1

Contents

Editor's Introduction

The papers in this special issue are all versions of presentations given at the conference on *Technology and the Rest of Culture* held at the New School in January of 1997. The conference was the fifth in a continuing series of Social Research conferences that began in 1989, all of the proceedings of which have appeared as special issues of this journal.

This series has several defining characteristics. First, it recognizes that even the most urgent contemporary events and issues are rarely unprecedented in human history, and second, that earlier experiences and ideas, despite their obvious capacity to illuminate the present, are frequently forgotten in the heat of public controversy. At the same time, this series recognizes that there is an unfortunate tendency to address difficult and complicated matters with narrow expertise, even when meaningful insights and possible answers may be found by bringing to bear a wider range of perspective and ideas.

We therefore have conceived our mission as one of fostering an ongoing public forum in which matters of grave interest and concern can be explored not only in terms of their immediate import, but also within their broad and rich historical and cultural context. More than reaffirming the maxim that to forget history is to risk repeating it, we seek through this effort to assure a more intellectually inclusive, reflective, and calmer understanding of the "hot" issues of the day.

To realize this mission, these conferences upon which these special issues are based bring together scholars and practitioners from a broad array of fields and disciplines, guaranteeing that topics will not be examined through a single lens, but rather from a virtual kaleidoscope of perspectives ranging from the historic, political, social, economic, and scientific to the philosophical, legal, and aesthetic. This intellectual breadth is reinforced through our collaborations with other cultural institutions that bring their own special expertise and resources to bear on our search for understanding. Thus none of our conferences have been stand-alone events. Instead, each has benefited from a collaboration between the New School, which is the venue of the actual conference, and major cultural institutions, which sponsor their own related exhibitions, lectures, readings, and other events, primarily in their own facilities.

We take pride in setting a new standard for such inter-institutional collaborations. For example, the first conference "In Time of Plague: The History and Social Consequences of Lethal Epidemic Diseases," examined the AIDS epidemic in light of the past, and was accompanied by an exhibition at the American Museum of Natural History. The next, "Home: A Place in the World," which explored the meanings of home and belonging in the context of present-day homelessness and the loss of homeland was a collaboration with five major New York City museums. The third in the series, "Rescue: The Paradoxes of Virtue," celebrated the founding of the University in Exile at the New School and the fourth, "In the Company of Animals," explored the present controversy over our proper relationship to other animals in partnership with four museums and the Academy of American Poets. Finally with "Technology and the Rest of Culture" we enjoyed the collaboration of the Metropolitan Museum of Art and the Environmental Simulation Laboratory and the Media Studies Program of the New School.

Since at least the 1970's there has been widespread discussion about the emerging information society. Our adoption of new forms of electronic interconnectedness is quickly creating a complex and changed environment in which we are bound to lead our lives. Not only is there radio and television, which now almost seem part of our birthright, if not part of nature, but we routinely shop, bank, receive our salaries, and pay our bills electronically. Our libraries are becoming virtual and museum artifacts are increasingly digitalized and available in cyberspace. E-mail and faxes link us almost instantaneously to distant corners of the world. There is more information available at the click of a mouse than we know what to do with, and this information is at once centralized and decentralized. Atomization and isolation progress in parallel with globalization and the homogenization of cultures. Traditional cultures are appropriated, reshaped, and disseminated while imagined cultures emerge in new forms. Virtual reality approaches the authority and authenticity of historical reality.

With all this come increasingly significant, but frequently unnoticed changes in how we define ourselves. As in the past, the consequences our technological choices are immense. As in the past the newest and dominant technologies have become our metaphors for society and identity. Our concepts of mind are a prime example. Where once we described thinking as wheels turning in our heads, we now speak of the mind as a computer and when things go awry we no longer speak of having a screw loose but of having malfunctioning circuits. We even worry about whether computers can be said to be conscious. And these are by no means superficial changes. Rather as Langdon Winner, one of the participants in this issue has written, "In choosing our terms, we express a vision of the world and name our deepest commitments."

Now it is not as if these extraordinary technological changes have gone unnoticed. On the contrary, news about them and their potential consequences are to be found everywhere. However, and this is the point, much of the discourse about our new "information" environment has been fraught by polar fantasies about the presumed opportunities and dangers they portend. Its advocates predict utopia and a technological sublime, while its critics warn of loss, inequity, and collapse. Where its advocates see a resurgence of participatory democracy, its detractors see increased vulnerability to manipulation by those who control access. And perhaps most troubling of all is that we proceed as if these changes are inevitable. Leo Marx, another speaker at the conference has written that, "By now most people in modernized societies have become habituated to the seeming power of advancing technology and its products to change the way we live. Indeed the steady growth of that power is just another self-evident feature of modern life that calls for no more comment than the human penchant for breathing." More than anything it is this assumption of inevitability and its associated lack of consciousness about the values, ideas, and meanings inherent in these changes that motivated this conference.

It's title, which is not, as you have undoubtedly noted, Technology and Culture but rather the more awkward "Technology and the Rest of Culture" is also at the heart of the matter. It is meant to call attention to the idea that technology is not a thing, as Professor Marx and others make clear. It is not something separate and apart from the rest of our culture with a life of its own. And as long as we fail to acknowledge this we will continue to endow it with agency and autonomy, which in turn will have profound, often unintended, moral, and political consequences.

The "Technology and the Rest of Culture" conference was made possible by the very generous support of The Gilman Foundation, The Rockefeller Foundation, The Engineering Foundation and Interval Research Corp. We are extremely grateful to each of these institutions.

Arien Mack
1997

The Concept of Technology: History, Definitions, and Critiques

How and why did the concept of technology emerge? What are its precursors? What are its different meanings? What are its critiques—and how and why did they emerge? How do changing technologies and technologic metaphors transform visions of the past, the present, and the future?

Introduction BY ROBERT HEILBRONER

PERHAPS nothing is so deeply educational—I mean more effective in revealing previously unnoticed aspects of our existence—than the discovery that the words we use often becloud the very things we are talking about. If I were to seek a common property of the three papers that follow, it would be precisely their educational brilliance in this regard. Robert McC. Adams, Leo Marx, and Langdon Winner make us realize that all too often *technology*—the term that seems so clarificatory when we employ it to explain various problems and properties of modern society—can be just such a beclouding term. Like all educational advances, this is both discomfiting and reassuring. Whatever else the readers of these papers will learn, I am certain they will never again use the term *technology* as a cure-all, a vade medcum, an open sesame.

More specifically, Robert McC. Adams alerts us to the manner in which technology, so frequently invoked as a great Archimedean lever, is in fact inextricably enmeshed in the very institutions and lifeways of the society to which its leverage is presumably being applied. When we have finished his paper, we will think twice before declaring that technology is the "cause" of, say, unemployment or alienation or whatever. "What we need to recognize," he tells us, "is that technology is intimately embedded in the entire field of forces constituting society at large." If it is the mover, it is also the moved.

From a differently oriented, more historically rooted approach, Leo Marx disabuses us of the "comforting theme" that technology refers to a linear advance by which "homo sapiens [has] acquired its unique power over nature." Tracing the changing popular conceptions of the term, Marx warns us that, although we can speak with some confidence and clarity about the nature of particular technologies, since we cannot identify the defining

3

properties of *technology,* the singular noun, the "concept of *tech-nology* becomes hazardous to the moral and political cogency of our thought." In a word, we cannot attribute to disembodied things—"machines"—the complex social rearrangements in which they play at best only a strategic, not an all-embracing role.

Last, but certainly not least, Langdon Winner forces us to recognize the hidden rationales that strongly affect utopian and dystopian visions of the future. In his examination of what is often passed off as the technologically determined shape of things to come, we discover how sociological and economic considerations enter into the conception of technology itself.

The three papers differ, and so some extent disagree, in what they see as the implications of the contemporary embodiment of technology in processes that generate information to a much greater degree than physical product, that tend to transform tasks once recognized as "skilled" or "demanding" into mere routines, and that seem to threaten our own self-understanding, self-esteem, and social involvement. I think it fair to characterize Adams as cautiously hopeful—technology is overwhelmingly "a reservoir of underutilized promises rather than threats"; Winner as perceiving it largely as the basis for misleading social expectations and explanations; and Marx as agnostic with respect to outcomes, cautionary with regard to expectations.

Hence no single view emerges from these papers, whether diagnostic or prognostic. I find that to be welcomed. The tensions, sometimes disagreements, that develop among those best equipped to penetrate to the core of concepts are not the least educational aspect of this delicate task. The papers that follow are elegant examples of just such tensions, although I would not say disagreements. One reads them, not alone with a sense of new understanding, but with an awareness of new indeterminacies. Technology is far more deep-rooted, many-sided, and far-reaching than we ordinarily think it to be. This is the all-important educational contribution that these essays convey.

Social Contexts of Technology

BY ROBERT McC. ADAMS

N OT long ago fellow contributor Leo Marx introduced my theme with a provocative question: "Why not start with the intuitively compelling idea that technology may be the truly distinctive feature of modernity?" (Marx, 1992, p. 407). But welcome as I find his suggestion, its popular acceptance is admittedly unlikely. Things today are felt by many to be moving too fast for modernity itself, although little but confusion yet accompanies the introduction of postmodernism as a fashionable substitute.

Marx's query did not imply that technology has become an impersonal, external force threatening to take control over our lives. To the contrary, his point—and mine—is to urge that we take a more active, wise, and deliberate part in directing technological transformations that are going on all around us toward ends of our own choosing. In spite of the widespread popular association of new technologies with uncertainty and risk, they are preponderantly a reservoir of underutilized promise rather than of inherent danger.[1]

Technology has a central—but preponderantly facilitative rather than initiating—role in sustaining a historically unprecedented standard of living that most of us have come to take for granted. It will be no less vital in attaining that standard, and making further improvements in it, for all humankind. But for such claims to be justified, the term must be applied to more than tools, techniques, and the skills to employ them. Those uncontroversial but narrow components of technology may be likened to a deep, slow ocean current of ultimately compelling force, but it is one that ordinarily we can neither recognize nor act upon (Braudel, 1973, p. ix).

From a less circumscribed point of view, what *is* technology?

Human society has always had to depend on what can be broadly characterized as primarily technological means. How else have we ever organized ourselves functionally and hierarchically, extended our mastery over the constraints and uncertainties of our environment, and improved our collective ability to set goals and move toward their attainment? New inventions and innovations are necessary but not sufficient steps in the pursuit of these objectives. Their importance depends on the extent to which they not only are adopted for the uses originally envisioned but find their way into further, previously unanticipated uses.

We need to recognize, in short, that technology is intimately embedded in the entire field of forces constituting society at large. The frame in which it is conceived, modified, and put to diverse uses before giving way to something better is in the end almost indistinguishable from the larger set of algorithms society is continually devising to carry out its objectives. In this much more interactive and inclusive sense, technology is better thought of as a set of partly distinct, partly overlapping *sociotechnical systems* whose precise content and boundaries are intentionally left ill-defined (Brooks, 1980, pp. 65–66).

Clearly, not all technological innovations are positive in their effects. Technological means are, after all, just as crucial for coercion and exploitation as for more benevolent purposes. Still, the balance is overwhelmingly positive. It is only on this basis that the endlessly reshaped mass of artifacts, techniques, designs, and organizational strategies, and the bodies of tacit as well as formal knowledge on which these all rest—in effect, the technological base of the social order—has grown, irregularly but cumulatively and without significant reversal, throughout the human career.

Two aspects of contemporary technology especially justify identifying it with modernity. The first is its ongoing, if never fully consummated, dependence on, but also contribution to, the remarkable progress of contemporary scientific inquiry. To be sure, there remains a distinction in fundamental orientation between the two. Discoveries in science focus primarily on the natural world, while their technological counterparts are largely directed toward a no less rapidly changing material and organizational world of our own design. But in methods, training, major features of their underlying theoretical frameworks, and even in the spheres of daily activity of individuals, the two have been more and more involved in a growing convergence and overlap.

Largely because technological innovation is tied to complex, expensive production processes and marketing uncertainties, it moves at a slower, more cautiously incremental pace than scientific discovery. Even in the highly competitive information industry, as IBM's then-chief scientist observed some years ago, technological development tends to be "much more evolutionary and much less revolutionary or breakthrough-oriented." Regularity and precision are also less uniformly characteristic of modern industry than of science. At IBM, "things are sufficiently complex that much is done by rule of thumb and not by precise knowledge. Many factors enter in; some of them are even cultural" (Gomory, 1983, pp. 576, 579).

Although the difference is becoming one of gradations rather than qualities, science for the most part also embodies a concern for more general levels of discovery and explanation. Like technology, it operates within a framework of somewhat conventionalized values, priorities, assumptions, goals, standards, and methods, all of them periodically subject to paradigmatic shifts (Kuhn, 1970). Situational determinants of the course of scientific advance are weaker, however, since the dominant concern is usually to make progress toward a generalized understanding of phenomena and not an immediate, practical outcome. But to apply an illuminating metaphor given by Harvey Brooks, while science and engineering can exist independently, like DNA strands, they are tending today to become meaningful only when they are paired (Brooks, 1994, p. 479).

In technology as well as science, the search for new innovations has become progressively more systematic and necessarily specialized. New institutions and incentives have been devised that explicitly systematize and stimulate the invention and adoption of new technologies. The lag between even the most basic scientific discoveries and their adaptation to practical ends has grown progressively shorter, and may presently almost disappear. Instruments for manipulating and measuring, always a source of creative, two-way feedbacks between science and technology, play an ever more central part in the advance of both. As the trained scientific-engineering community grows exponentially, specialization is carried to new extremes—but so also is the specialization of transdisciplinary searches for ideas and discoveries, leading to new forms of integration and application. A second major feature of our modern world with strong technological roots involves enormous, ongoing increases in international movement and communications. This development has led to a widening of human horizons that is surely welcome in most respects, even though the accompanying weakening and subordination of formerly more autonomous, local institutions and loyalties is a source of regret. More directly responsible than technology for the erosion of a locally constituted sense of order and security, in any case, is the competitively driven transformation of commerce and industry in an increasingly high-stakes pursuit of worldwide markets. Striking new advances in technology are indispensable in this process, but they are employed as means to ends for which the responsibility remains exclusively with human actors.

While technology is not a prime mover independently setting the course of wider societal developments, there are respects and times in which one can think of a kind of momentum being associated with technology. To begin with, over the very long run it has tended to prove unique among human cultural achievements in being cumulative and irreversible. Moreover, its improvement through time has not been constant or continuous but instead has taken the form of an irregular series of powerful thrusts or waves. At first occurring only at long intervals, these impulses have over time become much more frequent. Now, in fact, they may be almost continuous.

More directly suggestive of the existence of a kind of momentum in certain instances are relatively brief, more restricted technological trajectories. Consisting of fairly linear sequences of innovations, they proceed from a substantially original invention or discovery to many subsequent improvements in it and additional applications of it. The impressive further gains in the power and efficiency of steam engines that ensued over many subsequent decades from James Watt's original patent in 1769 are a classic example. Possibly the same could be said of Watt's engine itself, although it borrowed little more than its basic principles— and recognition of the growing need for rotary power—from its Newcomen forerunner more than a half-century earlier.

Let me briefly describe two earlier waves or clusters that may help in understanding the transformative process through which we are passing currently. The first episode of clustering that can be identified with any clarity was in some respects an antecedent Age of Information. One of the most salient features of what is often termed the Urban Revolution in ancient Mesopotamia was the first introduction of writing. The shift to urban life may not have been entirely voluntary on the part of most of the new city dwellers, for it is likely to have entailed higher exactions of produce and corvée labor from the populace on behalf of more authoritarian rulers than they had known previously. Once they were built, walled, and occupied, however, cities offered greater protection for food reserves than smaller, less defensible settlements in the dangerously variable, semiarid setting of the lower Mesopotamian plain. And with cities there arose an institutional framework for the exchange and redistribution of complementary resources between differently specialized eco-niches and their respective segments of the population—something that was beyond the scope of earlier, smaller communities.

Cuneiform writing on clay tablets had its origins toward the end of the fourth millennium B.C. Serving from the outset primarily the ends of economic administration in newly formed city-states, writing very quickly provided mnemonic aid on the contents of state or temple storehouses, regular distributions of rations, and other expenditures and receipts of critical resources. Over time, the originally pictographic forms were simplified and standardized into a mixed ideographic and syllabic system, finally beginning to approximate the grammatical structures of ordinary speech and serving the wider purposes of substantive communication. The transition from prehistory to history might be said to have been completed at this point, although in fact historical narratives in the usual sense still lay far in the future.

While there had been earlier, much less versatile and developed forms of symbolic notation, it cannot be accidental that the most decisive step in the first appearance of writing coincides with the first appearance of substantially larger population aggregates—on the order of at least several tens of thousands of people. City walls and monumental public buildings signify the emergence of underlying needs and forms of organizational complexity that were as new and unprecedented as were these massive architectural innovations.

Cuneiform signs testifying to the presence of status hierarchies and politicoreligious leadership are found in some of the earliest texts. So is evidence of an elaborate division of labor, including diverse groupings of craft specialists working in wood, metals, and more exotic raw materials that had to be brought from great distance, whose products exhibit the same extraordinary technical and stylistic precocity as the earliest tablets themselves (Nissen et al., 1993). In short, writing, crafts, and new forms of administrative organization emerged together, and rather abruptly. They appear to constitute a single complex, some combination of both stimulus and response to major changes in social scale and complexity. The developments of that era bear a generic resemblance, but no more, to the full onset of our own Information Age almost five millennia later.

Closer to the model of our own, engulfing wave of technological change, although with many major points of difference as well, was the British Industrial Revolution in the late eighteenth and early-to-mid-nineteenth centuries. It is the outstanding, canonical example of a long-sustained episode of relatively accelerated industrial development with only relatively minor accompanying political reverberations. In less than three generations craftsmanship was transformed into disciplined, steam-driven factories and Britain became (for a time) the "workshop of the world."

The primary driving forces of the Industrial Revolution were largely confined to a relatively small number of strategic regions and industries. Key inventions and innovations mechanizing the production of cotton textiles and iron were at its heart. Presently joining the textile and iron industries as the third leg of a powerful triad of change was the introduction of steam-powered railroads that quickly revolutionized inland transport. Within these sectors, impelled by insatiable world markets and buttressed by Britain's dominance in oceanic shipping, a number of indices of growth reached or exceeded two orders of magnitude.

Recent revisions, however, have significantly scaled back earlier estimates of annual growth rates. Apparently it was the consistency of the upturn climb rather than the steepness of its slope that constituted the principal revolutionary element.

What can be said, in a brief and admittedly superficial fashion, about the setting in which these changes originated? First, London, as the greatest city and port of Europe, was a powerful engine of change. It acted as a vital generator of both ideas for new industrial ventures and demand for new consumer products. Secondly, important changes in the agricultural sector had already been under way through much of the preceding century. Thus the subsistence needs of the great industrial towns to come were fortunately anticipated by rising productivity. Third, great numbers of former tenant villagers were driven by spreading enclosures of their former tenancies into the ranks of a potential urban industrial work force. Fourth, at the upper levels of the income pyramid considerable new landed and mercantile wealth entered the system. But while this deepened the capital pool available for industrial investments and heightened appetites for consumables, a flow of corresponding benefits to the industrial work force may have been delayed for as long as two generations. Fifth, entrepreneurs made their appearance, as Adam Smith was one of the first to notice. They combined access to this new wealth with an imaginative grasp of how to combine new mechanical processes with the still dimly recognized organizational potentials of the factory system.

Science, it should be noted, does not appear in this list. It contributed very little directly to any of the major technological breakthroughs before the middle of the nineteenth century. (A widely acknowledged aphorism in technological history is that science contributed less to the steam engine than the steam engine did to science.) In a more largely symbolic sense, on the other hand, scientific attitudes may have become increasingly influential as they tended to identify "progress" with systematic measurement and experiment. And increasing reliance on systematic measurement and experiment was, indeed, an essential step in entering the Machine Age.

Even taken all together, no one would claim that features like these *explain* the onset of the Industrial Revolution. But at least they help to direct current efforts to understand its origins and internal dynamics—efforts that will no doubt go on indefinitely.

With this lengthy but still very incomplete background, I now must focus on the most striking example of technological change unfolding in our own times—the wave of scientific discoveries and technological innovations comprising the material basis for our Information Age. Almost beyond the limits of credibility is the rapidity and extent of the transformation we are witnessing, although its long-sustained progress recalls that of the Industrial Revolution. These aspects especially deserve brief description before I turn to the complexity and synergism of the linkages between the technology of the Information Age and its wider social—indeed, global—setting.

The computing power available in individual machines has multiplied by something on the order of a *trillion* times in the short space of five decades. Eclipsing anything comparable in human history, a twenty percent annual rate of decline in the price of a given amount of computing power has been maintained steadily for over three decades. At one one-hundredth of one percent of what comparable computer power cost a quarter-century ago, the amount of it that now can be carried in a shirt pocket is probably comparable to what was available to NASA itself in first venturing to the moon. At that time there were perhaps fifty thousand computers in the world. Today, apart from computer chips and circuitry in uncounted consumer goods, there are 140 million (Kaufmann and Smarr, 1993; *The Economist*, 1996).

It is as processors and storers of information that computers have achieved this explosive penetration into virtually every conceivable human activity. But this leads us beyond computers, to the generating, transmission, and digestion of information. It has recently been estimated that there has been a million-fold increase in the global information network's carrying capacity in the last two decades. Its impacts, good and bad, on the worldwide diffusion of styles, ideas, modes of entertainment, business and scientific data, and much else are beyond calculation. Only as a by-product of such growth can the foreign exchange trading even be conceived at its present *daily* level of 1.3 trillion dollars.

A limited kind of determinate directionality or linearity must figure in such a record of growth. Computers create the need for as well as possibility of their own peripherals. Similarly, effective utilization of aggregate computing power as it doubles every eighteen months or so demands the concurrent development of enormously enlarged signal transmission capacity. Recursively involving a redirection of basic scientific inquiry in order to develop new means to attain this end, increased capacity is now being made possible by fiber-optic cables that can simultaneously transmit 1.5 million telephone conversations. Yet we must not exaggerate the inevitability of the overall course that information technology's growth has followed. There has seldom been a clear basis for prediction—certainly not with regard to timing, and in most cases also not to substance—about any of the major breakthroughs. The sense of linearity becomes apparent only retrospectively.

Participants, even well informed, centrally placed participants, have typically been unable to form an accurate estimate of the potential of the new technologies they are responsible for having developed. Unexpected convergences of forces not originally deemed relevant, and even accidents, have had a decisive influence on the later directions that new technologies actually followed. Thomas Watson, IBM's renowned senior executive during the epochal changes at the birth of the mainframe computer and the Information Age it assured, earlier foresaw no commercial demand for computers and believed that a single computer placed in operation in the firm's New York headquarters "could solve all the important scientific problems in the world involving scientific calculations." The Bell Laboratories management, similarly, was initially unwilling to pursue a patent for the laser, on the grounds that electromagnetic transmissions on optical frequencies were of no foreseeable importance for the Bell System's interests in communications. New technologies, it appears, seldom can be seen at the outset even by those best equipped to understand their potential except as supplements to existing technologies and the existing systems of which they are a part (Rosenberg, 1994, pp. 219–24).

Information is not a commodity in any ordinary sense. It may be costly to acquire, but in most circumstances is almost absurdly easy to replicate. It does not diminish in value with use and diffusion, and may even help to generate its own growth as multiple new uses are found for it. Unlike falling transport costs beginning in the past century, effects of the declining price of information on its production and consumption are very difficult to measure. With information so largely an intangible, its boundaries grow progressively more blurred.

> It is hard to exaggerate how far ahead of the American government Silicon Valley has moved. Even the statistics are a quagmire. Nobody in Washington has yet worked out how to calculate the output of a software developer, or into which employment category to put designers for the World Wide Web (at the moment, they go into "business services, not otherwise classified"). California's governor, Pete Wilson, recently complained that economic figures generated in Washington were "virtually irrelevant to California's new economy" (*The Economist,* 1997).

While notoriously subject to cycles of boom and bust, the information industry's underlying pattern of extraordinary growth is beyond question. Representing almost half the world market for computers, peripherals, and software, the United States now spends $282 billion annually on information technology hardware alone—more than on other, traditionally cyclical sectors such as autos and construction. With an increasing number of jobs as well as rising pay, it may well have accounted for as much as a fourth of the growth in real wages and incomes in 1996. And it differs strikingly from the other sectors in its greater resistance to the usual controls of Federal Reserve policy-makers:

> The tools of monetary policy—interest rates—do not work very well on high tech. In particular, it matters more whether compelling new products and technologies are available. When high-tech industries are riding a technology wave, they can power their way through slowdowns in the macroeconomy, no matter what the Fed does (Mandel, 1997, pp. 61, 68).

The Internet, with its huge and unrestricted access, is already well on the way to becoming a major feature of organizational and professional as well as public communication. Its positive impacts on the scale and character of global intercommunication are multiple: the erasure of political and geographic boundaries;

the detachment of cost from distance and to some extent even from volume; the readiness with which a rich variety of media such as graphics can be incorporated; and the openness to self-constituted group formation and collaboration without a central locus of control that might facilitate political or regulatory constraint. Even taken only at this time and individually, these steps toward absolutely unimpeded freedom of association and expression are of striking importance. In combination and over the long term, they are breathtaking.

The enormously positive aspects of these new features are at once apparent. But also worthy of our attention are some of the uncertainties and prospective costs or tradeoffs they may imply, even if these balancing considerations are of little interest to most enthusiasts. It is the so-called productivity paradox, for example, that the measurable economic returns to date on most of the huge, ongoing investments in information technology have remained disturbingly low. Not implausibly this can be attributed to the intangibility of some of the gains that are being made, as well as to the protracted learning that an unprecedentedly complex new technology requires. But for the present at least, the purely economic foundations of the case for our having been brought by information technology to the threshold of a new age of productivity growth remain rather shaky. Moreover, as large computational systems become embedded in organizational routines, other problems are certain to appear. With time, they will become obsolescent and increasingly difficult to maintain. As their original designers and installers gradually disappear from the scene, the software needed to improve their operation will in many cases need to be entirely reinvented.

In this setting, it is especially appropriate to consider what some of the impacts of an Information Age are likely to be on colleges and universities. Granting our inability to see very far into the future, one must assume that there will be a significant erosion of the identity and boundaries of educational institutions. Enthusiasts tell us that the Internet, in particular, seriously introduces the prospect of "asynchronous" learning. In the abstract, who could object to greater flexibility of individually scheduled self-instruction? But are universities merely places in which it has been economical and convenient to pour knowledge into geographic concentrations of individual student-receptacles, or centers whose insistence on creative collisions of view in a common setting is of the essence? Does the "anytime, anyplace" model apply with equal effectiveness to subjects differing in the length and rigor of preliminary studies required before one can come to grips with their intellectual substance? How will long-established, campus-based institutions accommodate to, and set the standards for, proliferating new forms of education-at-a-distance? Faced with such a dispersal, will the current and growing financial stresses on universities presently begin to diminish—or more likely get worse? If the half-life of information technologies continues to shorten, as seems likely, how long will legislatures and boards of trustees acquiesce in writing off escalating Information Age expenditures as consumables when they were originally sought and approved as capital investments? Particularly at risk may be university libraries, for it is difficult to see how any transition to a new world of electronically accessed knowledge will quickly eliminate the need for parallel systems devoted to preserving in their existing hard copies the knowledge of the past. Can university presses survive, with their typically small print-runs of valuable publications for limited audiences? (Okerson, 1996). Most of us will welcome the appearance of diverse new arenas of discourse that promise a wider and perhaps more rapid dissemination of knowledge. But what will replace the measures to assure basic competence and accuracy that disciplinary bodies and peer-reviewed publications presently provide?

From another point of view, we find ourselves in a period of growing ambiguity and potential conflict over intellectual property rights as formally expressed in international copyright agreements. Recognition of priority of authorship and discovery is the basic motivation in an academic merit system—often substituting for more pecuniary rewards, if we are honest about it. Can it survive a state of affairs in which someone's scholarly work can be freely scanned and uploaded by someone else for general Internet dissemination?

Clearly in jeopardy, in the United States more than any other country, are core copyright industries (including films, records, and music scores as well as printed publications). Presently, they contribute more that $200 billion to the annual U.S. gross domestic product. But as a second-order consequence, the transparency of all international boundaries to advanced communications networks also promises to raise a host of thorny issues to disrupt international political and cultural relations. Governments can and will act on behalf of the special values of their peoples and societies—for different degrees of personal privacy or protection of the rights of consumers, for example, or for maintenance of unique languages, cultures, and political and ethical norms. It would be most unwise simply to assume that all such values can be ignored or swept away without real difficulty and costs of some kind. Thus, while it may be enormously positive in most of its impacts, the advent of a new era in the uses and dissemination of information presents us with not a few significant risks and uncertainties; the larger promise of what is to come does not allow us to neglect them.

A number of other trends of a primarily economic character, already well under way, seem certain to continue. Intensified global competition invokes a fluid, unsettling mix of corporate strategies. Mergers, leveraged buyouts, plant and product diversification, and cross-licensing can alternate quickly, with or without concurrent attempts to lock in global markets through massive up-front investment in research and development, automation, and marketing. Shortened product-development cycles and proliferating new consumer products and options may benefit consumers as a result. "70% of the computer industry's revenues come from products that did not exist two years ago" (*The Economist*, 1996, p. 10). But not all product differentiation in this or other fields meets real needs or is productivity-enhancing. Impressive cost reductions will continue in many parts of the computer industry, but its widening array of new applications will surely come at the cost of eroding employment security for many.

Information-based growth in manufacturing and service industries may ultimately create as many or more new employment opportunities as it terminates in the short run. Expert economic opinion on this question is divided, at least leaving grounds for some optimism. But in the long run, as Keynes famously observed, we will all be dead anyhow. Particularly at risk in *this* life are minorities; urban centers with aging infrastructure whose inferior schools have failed to enhance the productive capacities of their labor force; and older industries that rely primarily on labor for their value-added contributions. Beyond the acknowledged migration of low-skilled job concentrations to less developed countries, there is a growing trend toward out-sourcing even skilled information processing overseas. Coinciding with this is a reciprocal movement of trained software personnel from other countries—an estimated twelve to fifteen thousand of India's annual crop of fifty thousand information technology graduates, for example—on immigration visas or term contracts into U.S. industry (*The Economist,* 1997, p. 6). This is a race in which the outcome of complex, contrasting flows of human and information resources defies prediction.

A concave income pyramid is developing that increases in height while widening ominously at the base. At the upper end of the scale, university-educated professionals as well as the independently wealthy will assuredly continue to benefit from steeply increasing income stratification. Primarily to their advantage, financial markets will be an important beneficiary of the Information Age. Almost indefinite increases are in prospect for the speed, volume, worldwide range, orderliness, transparency, and efficiency of financial transactions. But while unquestionably valuable, these consequences of the Information Age seem more likely to intensify than to diminish the social problems of our times.

And the pattern of increasing stratification of wealth unfortunately applies not only within fully industrial countries (preeminently the United States) but between the developed and the least-developed nations. Moreover, technological advances in this area will do little to reduce the likelihood of future business cycles. As the economic preponderance of the information industry grows, its own volatility may, on the contrary, make them deeper or more frequent.

As all these trends continue, we are also witnessing the obsolescence and demolition of an older generation of giant plants geared to mass production. This regrettably extends to some of the great industrial cities that grew up around them. There is also a blurring of publicly measurable performance criteria, and an erosion of control by even the most powerful nation-states over international corporate activities. Former employment levels and security of employment are among the first and most common of national interests to suffer. With gains in global output not matched by comparable increases in demand, it is difficult to see how the process can be reversed.

Double-digit unemployment in much of western Europe leaves a significant part of the population dependent on state subsidies at levels that few would argue can be indefinitely maintained. The U.S. counterpart is improved by the more rapid growth of the service sector and by large numbers of new, small-firm start-ups, a natural form of rejuvenation that ultimately will be beneficial. But even in a vigorous sector like the computer industry, current job replacements are overmatched by short- and medium-term job losses that have accompanied downsizing. The overall effect is the gradual creation of a two-tier labor force, and the progressive de-skilling of one of its major components. The higher technical and organizational requirements of automation and lean production lead to an increasing dependence on trained scientific and engineering personnel. For those who qualify, the news is good; average pay in the Silicon Valley software industry is three times the U.S. overall average. But this number is in any case relatively small. It cannot fully compensate for the disruptive social impacts suffered by the work force at large.

This is the perplexing, and certainly challenging, new world we have irreversibly entered—a world of sharp contradictions, marked by intensified competition, erosion and insecurity of industrial employment, and growing stratification of incomes. It is hard to escape the sense that the present dominance of markets, under these conditions, needs to be balanced with other mechanisms and goals if we are to assure the preservation of a civilized society. But on the other hand, it is also hard to deny that markets have powerfully stimulated, in a way that nothing else we know could have even approached, a vast impulse of technological transformation that has enriched our lives. The information revolution that is taking place all around us both embodies and symbolizes this huge advance in our horizons of experience, attainment, and understanding, and in our opportunities to think and act freely, independently, and globally.

In itself, technology has not been the source of either the ensuing benefits or the accompanying risks and uncertainties. Yet as perhaps the only area of human activity that can produce indefinite growth and improvement, I believe that technology offers the most realistic means of finding our way through difficult problems that certainly lie ahead. As they have always been, those problems will continue to be largely of our own making.

Note

1 While having a different focus, this discussion partly parallels and extensively draws upon the author's fuller, recent study of a related set of issues (Adams, 1996).

References

Adams, Robert McC., *Paths of Fire: An Anthropologist's Inquiry into Western Technology* (Princeton: Princeton University Press, 1996).

Braudel, Fernand, *Capitalism and Material Life, 1400–1800* (New York: Harper Colophon, 1973).

Brooks, Harvey, "Technology, Evolution, and Purpose," *Daedalus* 109:1 (1980): 65–81.

Brooks, Harvey, "The Relationship between Science and Technology," *Research Policy* 23 (1994): 477–86.

Economist, The, "The World Economy: The Hitchhiker's Guide to Cybernomics" (September 28, 1996), Special Supplement.

Economist, The, "A Survey of Silicon Valley" (March 29, 1997), Special Supplement.

Gomory, Ralph, "Technology Development," *Science* 220 (1983): 576–80.

Kaufmann, William J. III, and Smarr, Larry J., *Supercomputing and the Transformation of Science,* Scientific American Library 43 (New York: W. H. Freeman, 1993).

Kuhn, Thomas S., *The Structure of Scientific Revolutions,* 2nd ed. (Chicago: University of Chicago Press, 1970).

Mandel, Michael J., "The New Business Cycle," *Business Week* (March 31, 1997): 57–68.

Marx, Leo, "Letter to the Editor," *Technology and Culture* 33 (1992): 407.

Nissen, Hans J. et al., *Archaic Bookkeeping: Early Writing and the Techniques of Economic Administration in the Ancient Near East* (Chicago: University of Chicago Press, 1993).

Okerson, Ann, "Who Owns Digital Works?" *Scientific American* (July 1996): 80–84.

Technology: The Emergence of a Hazardous Concept*

BY LEO MARX

> ". . . the essence of technology is by no
> means anything technological."
> —Heidegger[1]

New Concepts as Historical Markers

T HE history of technology is one of those subjects that most of us know more about than we realize. Long before the universities recognized it as a specialized field of scholarly inquiry, American public schools were routinely disseminating a sketchy outline of that history to a large segment of the population. They taught us about James Watt and the steam engine, Eli Whitney and the cotton gin, and about other great inventors and their inventions, but more important, they led us to believe that technological innovation is a—probably *the*—major driving force of human history. The theme was omnipresent in my childhood experience. I met it in the graphic charts and illustrations in my copy of *The Book of Knowledge,* a children's encyclopedia, and in the alluring dioramas of early Man in the New York Museum of Natural History. These

* An early version of this essay was the Richmond Lecture at Williams College, September 26, 1996. I am grateful to Robert Dalzell, Michael Fischer, Michael Gilmore, Rebecca Herzig, Carl Kaysen, Kenneth Keniston, Lucy Marx, David Mindell, George O'Har, Harriet Ritvo, Merritt Roe Smith, Judith Spitzer, and G.R. Stange for their helpful comments and criticism.

exhibits displayed the linear advance of humanity as a series of transformations, chiefly represented by particular inventions—from primitive tools to complex machines—by means of which Homo sapiens acquired its unique power over nature. This comforting theme remains popular today, and it insinuates itself into every kind of historical narrative. Here, for example, is a passage from a recent anthropological study of apes and the origins of human violence:

> Our own ancestors from this line [of woodland apes] began shaping stone tools and relying much more consistently on meat around 2 million years ago. They tamed fire perhaps 1.5 million years ago. They developed human language at some unknown later time, perhaps 150,000 years ago. They invented agriculture 10,000 years ago. They made gunpowder around 1,000 years ago, and motor vehicles a century ago (Wrangham and Peterson, 1996, p. 61).

This capsule history of human development from stone tools to Ford cars illustrates the shared "scientific" understanding, circa 1997, of the history of technology. But one arresting if infrequently noted aspect of this familiar account is the belated emergence of the word used to name the very rubric—the kind of thing—that allegedly drives our history: *technology*. The fact is that during all but the very last few seconds, as it were, of the ten millennia of recorded human history encapsulated in this passage, the concept of *technology*—in our sense of its meaning—did not exist. The word, based on the Greek root, *techne* (meaning, or pertaining to, art, craft) originally came into English in the seventeenth century, but it then referred to a kind of learning, discourse, or treatise, concerned with the mechanic arts. At the time of the Industrial Revolution, and through most of the nineteenth century, the word *technology* primarily referred to a kind of book; except for a few lexical pioneers, it was not until the turn of this century that sophisticated writers like Thorstein Veblen began to use the word to mean the mechanic arts collectively. But that sense of the word did not gain wide currency until after

World War I.[2] (It is curious that many humanist scholars—I include myself—have so casually projected the idea back into the past, and into cultures, in which it was unknown.) The fact is that this key word—designator of a pivotal concept in contemporary discourse—is itself a surprisingly recent innovation.

Why does that matter? From a cultural historian's viewpoint, the emergence of such a crucial term—whether a newly coined word or an old word invested with radically new meaning—often is a marker of far-reaching developments in society and culture. Recall, for example, Tocqueville's tacit admission, in *Democracy in America,* that he could not do justice to his subject without coining the strange new term "individualism" (Tocqueville, 1946, II, p. 98); or Raymond Williams, who famously discovered, in writing *Culture and Society,* a curious interdependence, indeterminacy, or reflexivity in the relation between concurrent changes in language and in society. Williams had set out to examine the transformation of culture coincident with the rise of industrial capitalism in Britain, but he found that the word *culture* itself, like such other key words as *class, industry, democracy, art,* had acquired its meanings in response to the very changes he proposed to analyze. It was not simply that the word *culture* had been influenced by those changes, but that its meaning had in large measure been entangled with—and in some degree generated by—them (Williams, 1983, pp. xiii–xviii). A recognition of this circular process helps to account for the origin—and the significance—of *technology* as a historical marker.

But how do we identify the changes in society and culture marked by the emergence of *technology?* I assume that those changes in effect created a semantic void, that is, a set of social circumstances for which no adequate concept was yet available—a void that the new concept, *technology,* eventually would fill. It would prove to be a more adequate, apt referent for those novel circumstances than its immediate precursors—words like *machine, invention, improvement,* and, above all, the ruling concept of the *mechanic* (or *useful* or *practical* or *industrial) arts.* In a seminal essay of 1829, Thomas Carlyle had announced that the appropriate name for the emerging era was "The Age of Machinery" (Carlyle, 1829). But later in the century, *machinery* evidently came to seem inadequate, and the need for a more apt term evidently was felt. The obvious questions, then, are: Why was there a semantic void? Which new developments created it? What meanings was *technology* better able to convey than its precursors? In trying to answer these questions, I also propose to assess the relative merits and limitations of the concept of *technology*.

As for the *hazardous* character of the concept, at this point I need only say that the alleged hazard is discursive, not physical. I am not thinking about weaponry, nor am I thinking about the destructive uses of other technologies; rather, I have in mind hazards inherent in, or encouraged by, the concept itself—especially when the singular noun (*technology*) is the subject of an active verb, and thus by implication an autonomous agent capable of determining the course of events, as we constantly hear in countless variants of the archetypal sentence: "Technology is changing the way we live." When used in this way, I submit, the concept of *technology* becomes hazardous to the moral and political cogency of our thought. My argument, let me add, should not—if sufficiently clear—provide comfort to either the luddites or the technocrats. On the contrary, my hope is that it may help to end the banal, increasingly futile debate between these two dogmatic, seemingly irrepressible parties.

The Mechanic Arts and the Changing Ideology of Progress

By the 1840s, some of the changes that contributed to the emergence of the concept of *technology* were becoming apparent. They may be divided into two large categories, ideological and substantive: first, changes in the prevailing ideas about the mechanic arts, and second, changes in the organizational and material matrix of the mechanic arts.[3] As a reference point for both kinds of change, here is the peroration of a ceremonial speech delivered by Senator Daniel Webster at the opening of a new section of the Northern Railroad in Lebanon, New Hampshire, on November 17, 1847:

> It is an extraordinary era in which we live. It is altogether new. The world has seen nothing like it before. I will not pretend, no one can pretend, to discern the end; but every body knows that the age is remarkable for scientific research into the heavens, the earth, and what is beneath the earth; and perhaps more remarkable still for the application of this scientific research to the pursuits of life. The ancients saw nothing like it. The moderns have seen nothing like it till the present generation. . . . We see the ocean navigated and the solid land traversed by steam power, and intelligence communicated by electricity. Truly this is almost a miraculous era. What is before us no one can say, what is upon us no one can hardly realize. The progress of the age has almost outstripped human belief; the future is known only to Omniscience.[4]

The first ideological development that the word *technology* would eventually ratify, as indicated by Webster's exemplary display of the "rhetoric of the technological sublime," has to do with the perceived relation between innovations in science, the mechanic arts, and the prevailing belief in progress.[5] When Webster depicts the railroad as epitomizing—indeed, as constituting in itself—the progress of the age, he is confirming a subtle modification of the earlier Enlightenment concept of history as a record of progress.

Of course the idea of progress had been closely bound up, from its inception, with the accelerating rate of scientific and mechanical innovation. To call progress "an idea," incidentally, as if it were merely one idea among many, is to belittle it. By the time of Webster's speech, it had become the fulcrum of an all-encompassing secular world view, and, in a sense, modernity's nearest secular equivalent of the creation myths that embody the belief systems of premodern cultures. In the context of the seventeenth-century scientific revolution, the word *progress* had served, in a straightforward literal sense, to signify a series of incremental advances, within clearly bounded enterprises with specific goals, such as the development of the microscope or tele-scope. But later, in the era of the American and French revolu-tions, so many examples of this once clearly defined and bounded kind of progress had become manifest that the word's meaning was extended to the entire—boundless—course of human events. History itself was redefined as a record of the steady, cumulative, continuous expansion of human knowledge of, and power over, nature—knowledge and power that might be expected to result in a universal improvement in the conditions of human life.

But the republican thinkers who led the way in framing this "master narrative" of progress—men like Condorcet and Turgot, Paine and Priestley, Franklin and Jefferson—did not, like Webster, equate progress with innovations in the mechanic arts. They were radical republicans, political revolutionists, and although they celebrated innovations in the mechanic arts, they celebrated them not as constituting progress in themselves, but rather as the means of attaining it; to them the true measure of progress was to be humanity's forthcoming liberation from aristocratic, ecclesias-tical, and monarchic oppression, and the establishment of more just, peaceful societies based on the consent of the governed. What requires emphasis here is their strong conviction about the relationship between the arts and the rest of society and culture. To them, advances in science and the mechanic arts were chiefly important as a *means* of arriving at social and political *ends*.[6]

By Webster's time, however, that distinction already had lost most of its force. This was partly due to the presumed success of the republican revolutions, and to the complacent conservatism induced by the rapid growth of the immensely productive and lucrative capitalist system of manufactures. Thus Senator Webster, whose most important constituents were factory owners, mer-chants, and financiers, did not think of the railroad as merely instrumental—merely a means of achieving social and political progress. He identified his own interests with the company direc-tors and stockholders who enjoyed the profits, and in his view the railroad exemplified a socially transformative power of such immense scope and promise as to be a virtual embodiment—the perfect icon—of progress.

The new entrepreneurial elite for whom Webster spoke was thus relieved of its presumed obligation to fulfill the old republican political mandate. Although the Boston Associates—the merchants who launched the Lowell textile industry—were concerned about the social and political effects of their new industrial venture, they chiefly expressed their sense of social obligation by acts of private philanthropy (Dalzell, 1987). Innovations in the mechanic arts could be relied upon, in the longer term, to issue in progress and prosperity for all. A distinctive feature of the new mechanic arts, moreover, was their tangibility—their omnipresence as physical, visible, sensibly accessible objects. Thus new factories and machines might be expected, in the ordinary course of their operations, to automatically disseminate the belief in progress to all levels of the population. As John Stuart Mill acutely observed, the mere sight of a potent machine like the railroad in the landscape wordlessly inculcated the notion that the present is an improvement on the past, and that the wondrous future is imaginable, as Webster put it, "only to Omniscience" (Mill, 1865, II, p. 148).

But in the 1840s this blurring of the distinction between mechanical means and political ends also provoked ardent criticism. It was denounced by a vocal minority of dissident intellectuals as a sign of moral negligence and political regression. Thus Henry Thoreau, who was conducting his experiment at the pond in 1847, the year Webster gave his speech, writes in *Walden*:

> There is an illusion about . . . [modern improvements]; there is not always a positive advance. . . . Our inventions are wont to be pretty toys, which distract our attention from serious things. They are but improved means to an unimproved end (Thoreau, 1950, p. 46).

And in *Moby-Dick*, Herman Melville, after paying tribute to Captain Ahab's natural intellect and his mastery of the art of whaling, has him acknowledge the hazardous mismatch between his technical proficiency and his irrational purpose: "Now, in his heart, Ahab had some glimpse of this, namely, all my means are sane, my motive and my object mad" (Melville, 1967, p. 161).

This critical view of the new industrial arts marked the rise of an adversary culture that would reject the dominant faith in the advance of the mechanic arts as a self-justifying social goal. Indeed, a direct line of influence is traceable from the intellectual dissidents of the 1840s to the widespread 1960s rebellion against established institutions—from, for example, Thoreau's recommendation, in "Civil Disobedience" (1849); to "Let your life be a counter-friction to stop the machine" (Thoreau, 1950, p. 644); to Mario Savio's 1964 exhortation to Berkeley students: "You've got to put your bodies upon the [machine] and make it stop!" (Lipset and Wolin, 1965, p. 163). From its inception, the countercultural movement of the 1960s was seen—and saw itself—as a revolt against an increasingly "technocratic society."[7]

The Construction of Complex Sociotechnological Systems

I turn now to the substantive changes in the material character and organizational matrix of the mechanic arts that also helped to create the void to be filled by the new concept of *technology*. In Webster's view, these changes were embodied in the new machine itself—the railroad as a material and social artifact. Early in the industrial revolution innovations in the mechanic arts had been typified by single, freestanding, more or less self-contained mechanical inventions: the spinning jenny, the power loom, the steam engine, the steamboat, the locomotive, the dynamo, or, in a word, machines. By Webster's time, however, the discrete machine was replaced, as the typical embodiment of the new power, by a new kind of sociotechnological system. The railroad was one of the earliest, most visible of these large-scale, complex systems in the modern era.[8] A novel feature of these systems is that the crucial physical-artifactual, or mechanical component—

the steam locomotive, for example—constitutes a relatively small part of the whole.

Thus, in addition to the engine itself, the operation of a railroad required: (1) various kinds of ancillary equipment (rolling stock, stations, yards, bridges, tunnels, viaducts, signal systems, and a huge network of tracks); (2) a corporate business organization with a large capital investment; (3) specialized forms of technical knowledge (railroad engineering, telegraphy); (4) a specially trained work force with unique railroading skills, including civil and locomotive engineers, firemen, telegraphers, brakemen, conductors—a work force large and resourceful enough to keep the system going day and night, in all kinds of weather, 365 days a year; and (5) various facilitating institutional changes, such as laws establishing standardized track gauges and a national system of standardized time zones.

Eventually these large, tightly organized yet amorphous networks—like the telegraph and wireless systems, the electric power and use system, and so on—led to the replacement of the traditional family (father and sons) firm by the corporation as the dominant American form of business organization, and to the emergence of a new kind of professional or (as it later would be called in the United States) "scientific" management (Bijker et al., 1987, pp. 51–82; Chandler, 1977, pp. 79–120). A prominent feature of these complex, messy, ad hoc systems is the lack of clear boundary lines between their constituent elements. Of central significance here is the blurring of the boundary between the material-artifactual component (the mechanical equipment or hardware) and the rest: the cognitive, technical, or scientific components; the hierarchically organized work force; the financial apparatus; and the method of obtaining raw materials.

Another development that contributed to the complexity, scale, and singularity of the new systems was the increasing convergence, in the nineteenth century, of scientific knowledge and the mechanic arts. Webster had alluded to electricity and the telegraph, and had linked the advent of the railroad to "scientific research into the heavens, the earth, and what is beneath the earth." The fact is that the building of the railroads did mark a new departure in this respect. Whereas most earlier innovations of the industrial revolution had been made by practical, mechanically adept, rule-of-thumb tinkerers with little or no scientific education, a number of West Point-trained military engineers brought a more formal kind of technical education, in part derived from the Ecole Polytechnique, to the building of the American railroads (hence the emergence of *civil* engineering, to distinguish the civilian from the military branches of the profession).[9] By 1847, the joining of science and the practical arts was under way, but it was not until the end of the century, with the growth of the electrical and chemical industries, that the large-scale amalgamation of science and industry helped to call forth the concept of a new realm of innovation and transformative power—a new entity—called *technology* (Noble, 1977).

As early as 1828, to be sure, the prospect of amalgamating science and industry already had elicited an explicit statement—evidently the first ever made—about the need for that new concept. In a series of lectures at Harvard entitled "The Elements of Technology," Jacob Bigelow, a Boston botanist and physician, put the case this way:

> There has probably never been an age in which the practical applications of science have employed so large a portion of talent and enterprise . . . as in the present. To embody . . . the various . . . [aspects] of such an undertaking, *I have adopted the general name of Technology, a word sufficiently expressive, which is found in some of the older dictionaries, and is beginning to be revived in the literature of practical men at the present day.* Under this title . . . [I will attempt] to include . . . the principles, processes, and nomenclatures of the more conspicuous arts, particularly those which involve applications of science, and which may be considered useful, by promoting the benefit of society, together with the emolument of those who pursue them.[10]

But Bigelow was far ahead of his time. The concept of *technology* did not gain currency in the intellectual world for almost a century. His greatest success in disseminating the new term probably was its precocious use in naming a new institution of learning— The Massachusetts Institute of Technology—in 1862. (He also became a trustee of MIT.) But even at the mid-century, few writers availed themselves of the term. Karl Marx and Arnold Toynbee (a forebear of the twentieth-century historian), both of whom wrote extensively about the changes effected by the new machine power, seldom if ever used it. As late as the first (1867) edition of *Capital*, where Marx's subject—the way "machinery . . . forms new systems of manifold machines"—cries out for the new concept, he relied on terms like *factory mechanism*, and other relics of the old mechanistic lexicon (Marx, Karl, 1978). At points in Toynbee's influential lectures on the Industrial Revolution, composed in 1880–81, where *technology* would have been apposite, he also relied on conventional older terms like *mechanical discoveries, improvements,* or *inventions.*[11]

Early in the twentieth century the avant-garde of the modernist movement in the arts, with its several technology-affirming submovements—including the vogue of "Machine Art" and of machine-like styles in Futurism, Precisionism, Constructivism, Cubism, and the International Style in architecture—helped to elevate motifs formerly treated as merely instrumental to the plane of intrinsic (verging on ultimate) aesthetic value. In the Bauhaus aesthetic, design was married to industry. Indeed, the entire modernist turn to Mondrian-like abstraction—the new respect accorded to novel geometric, rectilinear, nonrepresentational styles—comported with the markedly abstract, mathematical, cerebral, practical, artificial (that is, not "organic" or "natural") connotations of the emerging concept, *technology.*

But the word itself did not gain truly popular currency until well after the astonishing explosion of inventions in the decades (roughly 1880–1920) bracketing the turn of the century. That decisive period, sometimes called the Second Industrial Revolution, marked the advent of electric light and power, the automobile, the radio, the telephone, the airplane, and the moving picture. As compared with the innovations of the first industrial revolution, these inventions were marked by their relative cleanliness, and by their dependence on advances in science. Each of these artifacts eventually formed the material core of a large, complex sociotechnological system. Each also was sufficiently impressive for inclusion in the iconology of progress. Of all the enduring testimonials to the dynamism of that era, none conveyed a more vivid sense of the accelerating rate of change keyed to new inventions than *The Education of Henry Adams* (first published privately in 1907). Adams announced the appearance of a new American, "born since 1900," who was

> the child of incalculable coal-power, chemical power, electric power, and radiating energy, as well as new forces yet undetermined—[and who] must be a sort of God compared with any other former creation of nature. At the rate of progress since 1800, every American who lived to the year 2000 would know how to control unlimited power. He would think in complexities unimaginable to an earlier mind (Adams, 1973, pp. 496–97).

Adams rarely if ever used the term *technology*, and in retrospect indeed his preferred vocabulary—*energy, power, forces*—often seems more vivid and evocative, more effective rhetorically, than the new term. But in spite of—or perhaps because of—its lack of connotative resonance, *technology* began to take hold of the imagination of writers in the early years of the new century. By 1904, Thorstein Veblen, who perhaps did more than any of his contemporaries to popularize the idea of *technology* and its unique transformative power, asserted that "The factor in the modern situation that is alien to the ancient regime is the machine technology, with its many and wide ramifications." He contended that this radically innovative mode of making and doing would literally transform the mental processes of those who used it.

The machine compels a more or less unremitting attention to phenomena of an impersonal character and to sequences and correlations not dependent for their force upon human predilection nor created by habit or custom. The machine throws out anthropomorphic habits of thought. It compels the adaptation of the workman to his work, rather than the adaptation of the work to the workman. . . . [It] gives no insight into questions of good and evil, merit and demerit. . . . The machine technology takes no cognizance of . . . rules of precedence; . . . it can make no use of any of the attributes of worth. Its scheme of knowledge . . . is based on the laws of material causation, not on those of immemorial custom, authenticity, or authoritative enactment. Its metaphysical basis is the law of cause and effect, which in the thinking of its adepts has displaced even the law of sufficient reason (Veblen, 1932, pp. 303, 310–11).

Veblen, along with Frederick Winslow Taylor and Howard Scott—who led the Technocracy Movement of the 1930s—also helped to popularize the seductive idea, foreshadowed in Webster's 1847 speech, that the miraculous improvements in the conditions of life made possible by technology might enable society to dispense with politics as its primary means of directing social change. This line of thought may be said to have culminated in the "liberal consensus" of the Kennedy era, when enthusiasm for the power of technology to replace politics became the quasiofficial doctrine of the administration; it was accompanied by confident academic predictions of the forthcoming "end of ideology."

Technology *Fills the Void*

At the outset I suggested that Daniel Webster's 1847 speech points to the existence of a conceptual void that would eventually be filled by the idea of *technology*. What was missing, from an ideological standpoint, was the concept of a form of power—of progress — that far exceeded, in degree, scope, and scale, the relatively limited capacity of the merely *useful* (or *mechanic* or *practical* or *industrial*) *arts* to generate social change. What was needed was a concept that did not merely signify, like the useful arts, a means of achieving progress, but rather one that signified a discrete entity that, in itself, virtually constituted progress. Besides, the idea of utility had long borne the stamp of vulgarity. Ever since antiquity, the useful arts in their various guises, had been regarded as intellectually and socially inferior to the *high* (or *fine, creative,* or *imaginative*) arts. The concept of the *useful arts* and its variants implied, if only because it explicitly designated a subordinate branch of the all-inclusive entity, the arts, a limited and limiting category. Indeed, the distinction between the useful and the fine arts had served to ratify a set of invidious distinctions between things and ideas, the physical and the mental, the mundane and the ideal, body and soul, female and male, making and thinking, the work of enslaved men and that of free men. By associating the railroad with science, business, and wealth, Webster and his contemporaries created the need for a term that would erase this derogatory legacy and elevate the useful to a higher intellectual and social plane.

All of these ideological purposes, and more, were served by the relatively abstract, indeterminate, neutral, synthetic-sounding term *technology*. Whereas the *mechanic arts* called to mind men with soiled hands tinkering at workbenches, *technology* conjures up images of clean, well-educated, white male technicians in control booths gazing at dials, instrument panels, or computer monitors. And whereas the mechanic arts were thought of as belonging to the mundane world of everyday work, physicality, and practicality—of humdrum handicrafts and artisanal skills—*technology* is identified with the more elevated social and intellectual realm of the university. This abstract word, with its vivid blankness, its lack of a specific artifactual, tangible, sensuous referent, its aura of sanitized, bloodless cerebration and precision, helped to ease the introduction of the practical arts—especially the new engineering profession—into the precincts of the higher learning.

Turning to the other half of the conceptual void, what was missing, from an organizational and material standpoint, was a name for the novel entity—a distinct new kind of sociotechnical formation—which emerged in the nineteenth century. This new entity has been called "a large-scale technological system," but that term begs an important question: Which aspect is technological? Where, exactly, is the *technology*? To be sure, the indispensable material component of these formations invariably is a distinctive material device, a piece of equipment designed to facilitate production, transportation, communication, or for that matter any form of human making or doing. But as we have seen, over time that pivotal artifactual component had come to constitute an increasingly minute part of the whole. Think of the computer chip!

Although in common parlance nowadays this material aspect often is what the concept of *technology* tacitly refers to, such a limited meaning—as we saw in the case of the railroad—is ambiguous and misleading. It is ambiguous because, for one thing, the artifactual component only constitutes a part of the whole system, yet the rest is so inclusive, so various, and its boundaries so vague, that it resists being clearly designated. This ambiguity surely is a large part of what Heidegger had in mind when he enigmatically asserted that "the essence of technology is by no means anything technological," and it also goes, as we shall see, to my assertion that *technology*, as the concept is used in public discourse nowadays, is hazardous. For in the major contemporary technologies the material component—*technology* narrowly conceived as a physical device—is merely one part of a complex social and institutional matrix. In capitalist societies that matrix typically takes the form of a private corporation, bank, or public utility with a large capital investment. (It is of course relevant that the concept emerged at the end of the era characterized by what Alan Trachtenberg has called "the incorporation of America" [Trachtenberg, 1982]). But these large technological systems also may be embedded in other kinds of macroinstitution, for example, branches of government, such as the military or the space program, or universities. They typically include an organized body of technical know-how; a cadre of specially trained experts and workers; and a related university teaching and research program. Moreover, the functioning of these systems, or technologies, often entails the creation of special legislative and regulatory bodies, as well as ancillary organizations for the supply of raw materials and the distribution of its products.

There is a compelling logic implicit in the emergence of this ambiguous, unspecific, indeterminate, well-nigh indefinable concept, *technology*, as a name for these ambiguous, messy, incoherent, new formations. This congruence takes us back to Raymond Williams's insight into the curious interdependence, or reflexivity, involved in the social construction of historical markers like *culture* or—in this case—*technology*. Earlier, I noted the blurring of the lines of demarcation—internally as it were—between the various components of a particular mechanic art, and the reduced relative importance of the material-artifactual component. But even more significant, perhaps, is the breakdown of the boundary separating whole technologies from the rest of society and culture.

Consider, for example, automotive technology. Its defining, indispensable material core is of course the internal combustion engine, plus—naturally—the rest of the automobile chassis. But surely it also includes the mechanized assembly lines, the great factories, the skilled work force, the automotive engineers, the engineering knowledge, the corporate structures with their stockholders and their huge capital investments, and their networks of dealers and repair facilities. But where do we draw the boundary separating all of this from the rest of society and culture? Do we include, as part of automotive technology, the road-building and maintenance systems, the trucking industry, the indispensable feeder industries—glass, rubber, steel, aluminum, plastic, and so on? What about the mines that provide the raw materials? Indeed, what about the global oil industry—an offspring of automotive technology that vies, in size, wealth, and influence, with its parent? At its outer limits, the intricate interpenetration of automotive technology and the rest of society and culture seems boundless and, finally, indescribable. The economic role of automotive technology, in its most comprehensive sense, is incalculable; as a source of jobs, for example, it may well account for as much as a fifth of the American work force. To speak, as people often do, of the "impact" of a major technology like the automobile upon society makes little more sense, by now, than to speak of the impact of the bone structure on the human body. But it is when we speak of the overall impact of *technology*, when the term putatively represents a discrete category of human activity, that its most hazardous consequences come into view.

The Hazards of Reification

The hazardous character of the concept of *technology* is a direct consequence of the history just outlined. That history has two major strands. We encounter one strand in common parlance nowadays, when *technology* is used as if it referred to a tangible, determinate entity—a kind of thing. This usage is traceable to the word's tacit place in that familiar lineage of material artifacts, from stone tools to automobiles, introduced at the outset. Indeed, historians and other scholars in the human sciences now tend to project the concept of *technology* backward in time to encompass the entire history of tools. Yet, as we have seen, the concept only came into general use when, at the end of the nineteenth century, the age-old artifactual lexicon of the mechanic arts had become inadequate. That is where the second strand in the history of the term comes in. The idea of tools or machines or, for that matter, any other material artifacts did not begin to convey the complex, quasiscientific, corporate character of the new sociotechnical formations that emerged at that time. The curious fact is that the discursive triumph of the concept of *technology* is in large measure attributable to its vague, intangible, indeterminate character—the fact that it does *not* refer to anything as specific or tangible as a tool or machine. If the first strand gives us a concept of technology that overemphasizes the tangible, the second is so inclusive as to be amorphous. But then, we finally are compelled to ask, what sort of entity is technology? What does its history reveal about its essential nature?

A significant result of that history, with its unstable marriage of artifacts and socioeconomic structures, is that the concept, *technology*, is peculiarly susceptible to reification. To borrow George Lukacs's lucid definition, reification occurs when "a relation between people takes on the character of a thing and thus acquires a 'phantom-objectivity,' an autonomy that seems so strictly rational and all-embracing as to conceal every trace of its fundamental nature: the relation between people." A distinctive result of reification observed by Karl Marx, Lukacs reminds us, is the power exerted by commodities over human beings; in that case social relations between people were mysteriously endowed with an objective, even autonomous character (Lukacs, 1971, pp. 83–87). I believe that something similar has happened with technology, which also has taken on an objective character, as if it existed independent of its human creators, and is capable of controlling them by virtue of an autonomy alien to them (Winner, 1977).

But it will be said that, whatever its limitations, the concept *technology* remains indispensable as the name for an increasingly large portion of human activity at the end of the twentieth century. Witness its widespread use nowadays to convey a sense of the accelerating diffusion of new technologies; the rapidly expanding universe of gadgetry; the deepening involvement of innovative technologies in every imaginable aspect of contemporary life. Today, it is true, *technology* is the word we rely on to refer to each and all of these developments. It is a key word, in fact, in our discourse about the "new world order," with its global market organized around a technological armature of electronic communications. The commonplace is that the transformation of global society is being "driven" by the electronic revolution in technology (Smith and Marx, 1995).

The striking fact is, however, that the concept of *technology*, when invoked on this plane of generality, is almost completely vacuous. It rarely enables us to say anything of genuine interest or value, to attribute any characteristic applicable, across the board, to all or most technologies. It is impossible, for example, to say anything meaningful about the moral import of technology or technological innovation in general. We have long realized that some of our technologies are unequivocally evil, useful only for destruction, such as those used to produce nuclear bombs, land mines, or poison gas; and of course we also have unequivocally benign medical technologies, such as those capable of eliminating hitherto incurable diseases, or of performing unimaginably delicate, microscopic surgical procedures. Thus, technology, according to a banality most of us encountered as children, is capable of enhancing and destroying life; it is good and bad, and this inherent contradiction makes the futility of the unceasing debate between Luddites and Technocrats all too obvious. One reason that technology is hazardous, then, is that it stifles and obfuscates analytic thinking. When we try to explain why that is so, the answer points to the fact that we cannot say what the word means.

Earlier I asked, "What sort of entity is technology?" But the truth seems to be that it is not an entity at all. An entity, according to my dictionary, is something that exists as a particular and discrete unit. But technology, in the sense of the mechanic arts collectively, lacks both particularity and discreetness, and indeed it is no sort of unit whatever. This elusive nonentity cannot be identified with any particular kind of artifact, or any particular social group, profession, or institution; nor does it represent any specifiable body of ideas, methods, or principles. This semantic vacuity is tacitly confirmed by the apparent inability of philosophers to say exactly what they mean by *technology*. Definitions of the word have been notoriously unsatisfactory. Heidegger defines it chiefly by saying what it is not, and among the other influential attempts, perhaps the most frequently cited is that of Jacques Ellul, who locates *technology* in any manifestation of *technique*. By identifying it with every act of making or doing, material or social, he drains it of all particularity and discreetness; the result is that it has little or no useful, specifiable meaning (Ellul, 1964). The vacuity of the concept might not matter very much were it not for its omnipresence, and its implicitly portentous consequences. Today an immense chorus of intelligent people laments the fact that "we" (humanity), in the trite phrase, "do not know where technology is taking us."

The chief hazard attributable to the concept of *technology*, as currently used, is the mystification, passivity, and fatalism it helps to engender. Today we invoke the word as if it were a discrete entity, and thus a causative factor—if not *the* chief causal factor—in every conceivable development of modernity. Although we cannot say exactly what that "it" really is, it nonetheless serves as a surrogate agent, as well as a mask, for the human actors actually responsible for the developments in question. Because of its peculiar susceptibility to reification, to being endowed with the magical power of an autonomous entity, technology is a major contributant to that gathering sense, at the close of the millennium, of political impotence. By attributing autonomy and agency to technology, we make ourselves vulnerable to the feeling that our collective life in society is uncontrollable. The popularity of the belief that technology is the primary force shaping the postmodern world is a measure of our growing reliance on instrumental standards of judgment, and our corresponding neglect of moral and political standards, in making decisive choices about the direction of society. To expose this hazard is a vital task for the human sciences.

Notes

1 Heidegger, 1977, p. 4. For my earlier assessment of Heidegger's argument, see "On Heidegger's Conception of 'Technology' and Its Historical Validity" (1984).

2 The first use of the word in this sense reported by the *Oxford English Dictionary* was in 1859; but as noted below, Jacob Bigelow had used it as early as 1829, and it evidently had appeared in German, Swedish, French, and Spanish in the late-eighteenth century. Thus Johann Beck-

mann, a German professor, is credited by Siebicke (1968) and Gille (1986) with its first use in a book title, *Anleitung zur Technologie* (1777). See also Morere (XII, 1966). My version of this history, it should be said, is not based on the kind of comprehensive examination of primary sources that an authoritative account requires. Such a study, especially one that examines the history of the word in several modern languages, would be invaluable, but to the best of my knowledge, does not yet exist.

3 Although I rely on American examples, I believe that British and western European equivalents exist for many of them.

4 Webster, 1903, IV, pp. 105–107. For a more detailed analysis of the speech, see Leo Marx, 1964, pp. 209–14.

5 The "technological sublime" refers to the extension of the concept of sublimity, originally applied chiefly to the transcendent, quasitheological attributes of natural phenomena, to the new industrial artifacts. I discuss this tendency elsewhere (Marx, 1964, pp. 195–99); David Nye has made a comprehensive study of the subject (Nye, 1995).

6 Thus when Benjamin Franklin was offered a potentially lucrative patent for his ingenious new stove, he explained his refusal to accept by invoking the communitarian republican notion that inventions are valued for their contribution to the polity. "I declined it from a principle which has ever weighed with me on such occasions, that as we enjoy great advantages from the inventions of others, we should be glad of an opportunity to serve others by any invention of ours" (Franklin 1950, p. 132). For other discussions of this topic, see also Marx, 1987, and Marx, 1996.

7 Roszak, 1969, p. 5. Theodore Roszak, who helped to define the character of the student revolt, refers to the rebels as "technocracy's children," and "the technocracy" as "that social form in which an industrial society reaches the peak of its organizational integration. It is the ideal men usually have in mind when they speak of modernizing, updating, rationalizing, planning" (Roszak, 1969, p. 5).

8 I add the qualification, "the modern era," to acknowledge the provocative theory, advanced by Lewis Mumford (1966), that the first "machine" was in fact such a system, the systematic organization of work contrived by the Egyptians to build the pyramids. The trouble with this theory is that it ignores the artifactual component of the concept of the *machine* and, when it later emerges, the concept of *technology*. For a more extended analysis of this theory, see my essay, "Lewis Mumford, Prophet of Organicism" (1990, pp. 164–80).

9 Dunlavy, 1994; Hill, 1957. At West Point, the military engineers, trained in the tradition of the Ecole Polytechnic, acquired a more

sophisticated knowledge of geometry, physics, and of a general scientific viewpoint than most American engineers at the time. A number of them left the army and became "civil" engineers, and worked on the railroad. I am grateful to Merritt Roe Smith for calling my attention to this development.

10 Bigelow, 1829, pp. iii–iv (emphasis added). Bigelow's lectures were supported by the endowment of Count Rumford who, in his 1815 will, had left Harvard $1000 a year for lectures designed to teach the "utility of the physical and mathematical sciences for the improvement of the useful arts, and for the extension of the industry, prosperity, happiness and well-being of society" (Struik, 1948, pp. 169–70). Struik seems to have been the first historian to credit Bigelow with first using the modern sense of the word *technology.*

11 Marx, 1978, p. 403; Toynbee, 1960. Marx's discussion in *Capital,* I, Part IV, Ch. XV, "Machinery and Modern Industry," is of particular pertinence (Tucker, 1978, p. 403). As late as its eleventh (1911) edition, *The Encyclopaedia Britannica,* which contained no separate entry on *technology,* was offering the word *technology* as a possible alternative to the (preferred) use of *technical* in the entry on "Technical Education" (*The Encyclopaedia Britannica,* 11th ed., XXVI, p. 487).

References

Adams, Henry, *The Education of Henry Adams* (Boston: Houghton Mifflin, 1973).

Bigelow, Jacob, *Elements of Technology* (Boston: Boston Press, 1829).

Bijker, Wiebe E., Hughes, Thomas P., Pinch, Trevor J., eds., *The Social Construction of Technological Systems: New Directions in the Sociology and History of Technology* (Cambridge: MIT Press, 1987).

Carlyle, Thomas, "Signs of the Times," *Edinburgh Review* (1829), reprinted in *Critical and Miscellaneous Essays* (New York: Belford, Clarke & Co., n.d.), III, pp. 5-30.

Chandler, Alfred D., Jr., *The Visible Hand: The Managerial Revolution in American Business* (Cambridge: Harvard University Press, 1977).

Dalzell, Robert F., Jr., *Enterprising Elite: The Boston Associates and the World They Made* (Cambridge: Harvard University Press, 1987).

Dunlavy, Colleen, *Politics and Industrialization: Early Railroads in the United States and Prussia* (Princeton: Princeton University Press, 1957).

Ellul, Jacques, *The Technological Society* (New York: Knopf, 1964).

Franklin, Benjamin, *The Autobiography of Benjamin Franklin* (New York: Modern Library, 1950).

Gille, Bertrand, *The History of Techniques* (New York: 1986).

Heidegger, Martin, *The Question Concerning Technology and Other Essays,* William Lovitt, trans. (New York: Harper & Row, 1977).

Hill, Forest G., *Roads, Rails, and Waterways: The Army Engineers and Early Transportation* (Tulsa: University of Oklahoma Press, 1957).

Lipset, Seymour Martin, and Wolin, Sheldon, eds., *The Berkeley Student Revolt* (New York: Doubleday, 1965).

Lukacs, Georg, *History and Class Consciousness: Studies in Marxist Dialectics,* Rodney Livingstone, trans. (Cambridge: MIT Press, 1971).

Marx, Karl, *Capital* in Robert C. Tucker, ed., *The Marx-Engels Reader* (New York: Norton, 1978).

Marx, Leo, *The Machine in the Garden: Technology and the Pastoral Ideal in America* (New York: Oxford University Press, 1964).

Marx, Leo, "On Heidegger's Conception of 'Technology' and Its Historical Validity," *The Massachusetts Review* 25 (Winter 1984): 638–52.

Marx, Leo, "Does Improved Technology Mean Progress?" *Technology Review* (January 1987): 32–41.

Marx, Leo, "Lewis Mumford: Prophet of Organicism," in Thomas P. Hughes and Agatha C. Hughes, eds., *Lewis Mumford: Public Intellectual* (New York: Oxford University Press, 1990), pp. 164–80.

Marx, Leo, and Mazlish, Bruce, eds., *Progress: Fact or Illusion?* (Ann Arbor: University of Michigan Press, 1996).

Melville, Herman, *Moby-Dick* (New York: Norton, 1967).

Mill, John Stuart, "M. de Tocqueville on Democracy in America," *Edinburgh Review* (October 1840) in *Dissertations and Discussion; Political, Philosophical, and Historical* (Boston: William V. Spencer, 1865).

Morere, J.-E, "Les Vicissitudes du Sens de 'Technologie' Au Debut du XIXe Siecle," *Thales,* XII (1966): 73–84.

Mumford, Lewis, *The Myth of the Machine; Technics and Human Development* (New York: Harcourt, 1966).

Noble, David, *America by Design: Science, Technology, and the Rise of Corporate Capitalism* (New York: Knopf, 1977).

Nye, David E., *American Technological Sublime* (Cambridge: MIT Press, 1995).

Roszak, Theodore, *The Making of a Counter Culture: Reflections on the Technocratic Society and Its Youthful Opposition* (New York: Anchor, 1969).

Segal, Howard, *Technological Utopianism in American Culture* (Chicago: Chicago University Press, 1985).

Siebicke, Wilfried, *Technik, Versuch einer Geschicte der Wortfamilie um (techne) in Deutschland vom 16 Jahrhundert bis etwts a 1830* (Dusseldorf: 1986).

Smith, Merritt Roe, and Marx, Leo, eds., *Does Technology Drive History? The Dilemma of Technological Determinism* (Cambridge: MIT Press, 1995).

Struik, Dirk, *Yankee Science in the Making* (Boston: Little Brown, 1948).

Thoreau, Henry, *Walden and Other Writings* (New York: Modern Library, 1950).

Tocqueville, Alexis de, *Democracy in America*, Phillips Bradley, trans. (New York: Knopf, 1946).

Toynbee, Arnold, *The Industrial Revolution* (Boston: The Beacon Press, 1960.)

Trachtenberg, Alan, *The Incorporation of America* (New York: Hill & Wang, 1982).

Veblen, Thorstein, *The Theory of Business Enterprise* (New York: Scribners, 1932).

Webster, Daniel, *The Writings and Speeches of Daniel Webster* (Boston: Little, Brown, 1903).

Williams, Raymond, *Culture and Society, 1780–1950* (New York: Columbia University Press, 1983).

Winner, Langdon, *Autonomous Technology: Technology-out-of-Control as a Theme of Political Thought* (Cambridge: MIT Press, 1977).

Wranghan, Richard, and Peterson, Dale, *Demonic Males: Apes and the Origins of Human Violence* (New York: Houghton Mifflin, 1996).

Technology Today: Utopia or Dystopia?

BY LANGDON WINNER

ON the cathode-ray tube in front of her flashes an image of an envelope with a handwritten address—upside down. The woman touches the keyboard, flipping the picture 180 degrees. She reads the sloppy script and presses several more keys. The image vanishes and another appears in its place. The whole process has taken perhaps four seconds. Around her in seemingly endless rows of work stations sit other clerks, male and female, from their early twenties to early sixties, most with high school educations or better, silently staring at their screens, engaged in the same repetitive task: reading the addresses that the U.S. Postal Service computers cannot decipher, sending correct information back to distant post offices where bar codes will be attached.

"We handle no physical mail here at all," the plant manager proudly explains. "The images arrive over phone lines from central post offices in Connecticut, Massachusetts, and New York, are stored in four mainframe computers and then distributed to our 485 workstations. Yesterday we set a new record for this facility: 3,100,000 pieces processed in a single day, most of it Christmas cards in the seasonal rush."

The Remote Encoding Center, located in a large, windowless warehouse in Latham, New York, is fairly typical of work sites that now greet ordinary working Americans. Although operated by a government agency, the plant is organized by the logic of cost-cutting, technological dynamism, global communications, economic competitiveness, and flexible social relations that also characterizes private sector production in the late-twentieth century.

One of sixty-five such centers in the United States, it has been in operation for two years. Whether it exists for, say, another three years depends on how efficient it is as compared to its sister centers around the country. Since nothing about its work depends upon a specific geographical location, it could just as well process images of mail from Georgia or Oregon. By the same token, sites in the west and south could handle the work done in Latham. "So far we have an excellent record for productivity. Our employees know that unless we perform better than similar sites, these jobs will go elsewhere."

Eventually, the jobs will vanish in any case. As pattern recognition software incorporates the latest developments in artificial intelligence, the human contribution is quickly whittled away. Within a few years computers will be able to read the vast majority of the items the clerks now handle. Our guide assures us that the employees understand that even in the best of circumstances, their jobs have short horizons. Hiring policies at the center presuppose the ephemerality of the work. Of the eight hundred people the center employs, only a quarter are "career" postal workers with benefits and pensions. The rest are temporaries, slotted into four- to six-hour shifts. Welcome to the digital age.

The mood in the enormous work room is sober, bordering on grim. Although their activities depend on sophisticated communications equipment, there seems to be little communication among the clerks. Able to come and go as they please in pre-arranged, round-the-clock, flex-time schedules, they file into the building, stopping briefly at the coat room, clocking in and with plastic swipe cards, quietly taking a seat at a work station, logging into the computer. There is seldom any need to talk to another person all. In the brightly lit "cafeteria" in an adjoining room, there are only food vending machines with several tables and hardback chairs for those taking their five-minutes-per-hour break. Nothing about the space invites social gatherings or conversation.

As I wander among the workstations watching the clerks dispatch one image after another, I notice several cardboard cartons filled to the brim with AA batteries, discarded by workers, many of whom wear head phones connected to Walkman and Discman stereos. Each one listens to his/her own music—rock, rap, country, jazz, classical—to help the hours pass. There they sit, side by side, fingers gliding across the keyboard—alone together.

Digital Technology as Cultural Solvent

The visit to the Remote Encoding Center left me deeply conflicted. Should I be thankful that several hundred workers are earning decent wages—$12.00 an hour on average—in steady work that, back strain and carpel tunnel syndrome aside, most of them find fairly agreeable? Or should I yield to my basic instinct and recognize that when all is said and done this is numbing, stifling work, devoid of creativity, suppressing everything vital and interesting in the individual? Should I despair at the prospect that all of it will be electronic landfill within a decade? Or should I anticipate, as my engineering students always do, that much better jobs will be available in programming the next generation of machines?

For most people who think about the role of technological change in human well-being, such conflicts count for little. In our society the prevailing view stresses economic results to the exclusion of all other concerns. New instruments, techniques, and systems are seen as inputs that go into the hopper of material production and are mixed with other ingredients—education, marketing, government policy, and so forth. What comes out the other end is what businessmen, politicians, and economists uphold as most basic of social goods: economic growth. From this standpoint, we know we are doing well as a society when technological innovation contributes to a gradual increase in incomes, profits, stock prices, and living standards.

What the conventional view lacks, however, is any notion of technological development understood as a complex social, cultural, and political phenomenon. This is not an obscure or mysterious topic. Every thoroughgoing history of technological system-building points to the same conclusion, namely that technical innovations of any substantial extent involve a reweaving of the fabric of society, a reshaping of some of the roles, rules, and relationships that comprise our ways of living together. In this process, many people in many different situations contribute to the kinds of final outcomes we talk about as inventions or innovations. Of course, new material instruments and techniques are never the sole cause of the changes one sees. But the creation of new technical devices presents occasions around which the practices and relations of everyday life are powerfully redefined, the lived experiences of work, family, community, and personal identity, in short, of some of the basic cultural conditions that make us "who we are."

At some level most people appreciate that new technologies are involved in changing the practices and patterns of everyday life. But most find it difficult to talk about this, much less to move the social, political, cultural, and ethical considerations about technological transformation to the center stage of public debate. We realize that the technologies that surround us affect matters we deeply care about—the satisfactions of working life, the character of family ties, the safety and friendliness of local communities, the quality of our interactions with schools, clinics, banks, the media, and other institutions. But finding ways to deliberate, organize, or act on these intuitions is not part of our education or our competence.

Today the solely economic perspective on technological change seems increasingly hollow because it fails to illuminate forces and circumstances of crucial concern to a great number of people, both in our society and in many other parts of the world; forces and events starkly evident in places like the Remote Encoding

Center. On most lists, the key elements of this transformation include the following.

- To an increasing extent, the basis of wealth no longer depends, as it did in modern industrial economies, on access to material resources. Instead, wealth derives from applications of brainpower to the creation of marketable goods and services (Thurow, 1996).
- Productive operations now presuppose "global" extension in which the capital, information, expertise, and labor of any organization can be spread across several nations, regions, or continents and yet operate effectively in real time (Casells, 1996).
- Tried and true patterns of factory organization and corporate bureaucracy perfected throughout much of the twentieth century are being replaced by organizational principles of a much different sort. Organizations are reengineered to achieve flexibility, agility, and leanness, adapting their processes to suit closely the needs of their customers (Hammer and Champy, 1993).
- Because production facilities must turn on a dime, quickly altering what is produced and how, workers must be prepared for dramatic shifts in what they do. They too must be flexible and customer-oriented, employed within temporary teams with ever-changing objectives (Warme et al., 1992).
- Finally, one sees the continuing digital transformation of and astonishingly wide range of material artifacts and associated social practices. In one location after another, people are saying in effect: Let us take what exists now and restructure or replace it in digital format. Let's take the bank teller, the person sitting behind the counter with little scraps of paper and an adding machine and replace it with an ATM accessible twenty-four hours a day. Let's take analog recording and the vinyl LP and replace it with the compact disc in which music is encoded as a stream of digital bits. Or lets take the old-fashioned bookstore and transform it into a site on the

World Wide Web where people can read reviews and browse through the best-seller lists. The possibilities are limitless (Winner, 1996a, 1996b).

Taken together the transformations I have noted amount to a massive, ongoing social experiment whose eventual outcome no one fully comprehends. During the past two decades American workers have achieved steady increases in productivity, using a host of new tools and techniques. But except for incomes of perhaps the top 20 percent of the nation's populace, ordinary people have not seen the fruits of transformation return to them as an improved quality of life. In fact, during this period the average wages for Americans in the middle and lower levels of our society have declined. As innovations in computing, communication, and flexible production have multiplied, people see many of the jobs and workplaces that formerly sustained their way of life abruptly terminated or moved offshore. Whole categories of employment—telephone operators, accountants, secretaries, and many kinds of factory workers—have dwindled or expired. Those displaced from their former vocations often find that they must hold two or three jobs to maintain a middle-class standard of living (see Uchitelle and Kleinfeld, 1997). Most employees at the Remote Encoding Center, for example, take the work as a second job to supplement their incomes.

Of course, the consequences of such changes go far beyond the problem of sagging incomes. Even those who retain their jobs face anxiety about the immediate future. Recent polling data suggests that large numbers of people think the economy is "improving" nonetheless fear that they will soon lose their jobs (Lohr, 1996). To an increasing extent, the technological world of the late twentieth century is one that everyone is made to feel expendable.

Along the way many practices long associated with loyalty at work, stable families, and a sense of belonging in coherent communities no longer count for much. Sources of support for the institutions of civil society—schools, youth groups, churches, service clubs, and charitable organizations—dwindle because both employers and former employees have less time and money to give. As one of top five companies in the Fortune 500 withdraws from the city that was once its mainstay, it finds that it can no longer afford to sponsor a local little league team.

In this process many of the agreeable textures in the common life of earlier periods are eviscerated or placed under stress. In places formerly occupied by human beings and predicated on social interaction, we now find sophisticated hardware and software; the ATM and voice mail are notable examples. None of us can escape the influence of these systems, regardless of what we may think of them; for as we interact with these devices, our behaviors are automated as well. In America and other nations affected by globalism and the rise of a society based upon digital encoding, many of the roles, institutions, and expectations that were serviceable in previous decades are no longer welcome.

Technological Drivenness and Social Construction

For those who recognize how thoroughly our ways of living are intertwined with technical devices, evidence of change—even change that at first seems ominous—can be cause for hope. If it were possible to reflect upon and act intelligently upon patterns in technology as they affect everyday life, it might be possible to guide technocultural forms along paths that are humanly agreeable, socially just, and democratically chosen. For that reason, many intellectuals, activists, and artists now seize upon technological innovation as an arena in which humane, democratic conditions might be fostered. A good number, this writer included, are concerned to find approaches and strategies that might place technological choices at center stage.

The dominant perception of technology and culture, however, is not one that gives much credence to this hope. From every corner we are advised that far from broadening the options open to us, technological developments characteristic of the late twentieth century constrain our choice, forcing society in a particular direction, allowing no significant modification or appeal. One hears politicians, businessmen, and ordinary people exclaim that because technology in the late-twentieth century is moving in a particular direction, there is no alternative but to lay people off, close plants, dismantle institutions once crucial to the vitality of local culture, alter practices of schooling, adapt our patterns of family life, and so on. True, we may not have chosen the changes of our own volition. But we have no choice but to adapt and join defined by rapidly moving instrumentalities and organizational demands of today's high tech economy.

Thus, journalist Steward Brand exclaims in *Wired,* "Technology is rapidly accelerating and you have to keep up. Networks and markets, instead of staid old hierarchies, rule, and you have to keep up" (Brand, 1995, p. 38). Michael Hammer, leading advocate of "reengineering," surveys the social disruptions brought by the new "process-centered world," and concludes, "It is the inevitable result of technological advances and global market change. The question that we must confront is not whether to accept it but what we make of it" (Hammer, 1996, p. 265). In fact, the sheer drivenness of technological development is commonly held out to people as a bracing moral challenge, one that will test their character. Whether one succeeds or fails will depend upon how well one reads the powerful trajectory of social and organizational changes and positions oneself accordingly (Burrus, 1993).

Among sociologists and historians who study technological change, there is a jarring irony here. For within today's scholarly communities, once-popular notions of technological inevitability, determinism, and imperative have gone out of fashion. Three or four decades ago, debates about technology and society often focused on what were widely (but by no means uniformly) believed to be essential features of technology and technological change. Many economists, historians, and social theorists argued that the development and use of technology followed a fairly unilinear path, that technological change was a kind of univocal, determining force with a momentum and highly predictable outcomes.

There were optimistic and pessimistic versions of this notion. Among social scientists one influential group espoused what was called "modernization theory," the belief that all societies move through stages of growth, or stages of development linked to technological sophistication and social integration such that eventually they would reach what was called the "take-off point" and achieve the kind of material prosperity and way of life found in late-twentieth-century Europe and America—all to the good (see, for example, Rostow, 1960). There were also pessimistic variants of this conception, theories of technological society that focused upon the human and environmental costs of rapid technological development, for example the visions presented in Jacques Ellul's *The Technological Society* (1964), Herbert Marcuse's *One-Dimensional Man* (1964), and Lewis Mumford's *Myth of the Machine: The Pentagon of Power* (1970).

Whether taken in optimistic or pessimistic variants, there was something of an agreement that modern technology had certain essential qualities, among which one could list a particular kind of rationality—instrumental rationality, the relentless search for efficiency—and a kind of historical momentum with indelible features that rendered other kinds of social and cultural influences upon the character of social life far less potent.

During the past twenty-five years there has been an enormous effort to show that the idea that modern technology is a unilinear, univocal force is completely erroneous. Instrumental devices, systems, and techniques as well as the ways in which they are used and interpreted are always subject to complicated "social shaping" or "social construction."[1] Looking closely at how technologies arise and how they are affected by the contexts that contain them, one does not find a juggernaut foreordained to achieve a particular shape and to have particular consequences, but rather a set of options open to choice and a variety of contests over which choices will be made. Debunking work of that kind has been undertaken by European and American social scientists, historians, and philosophers. One purpose of this work is simply to provide a more faithful account of how technological innovation and associated social change actually occurs. Another goal is to snatch human choice from the jaws of necessity, to redeem the technological prospect from both the facile optimism of liberal, enlightened thought and the pessimism of cultural critics. Hence, an endless array of case studies and social theories now proudly affirm voluntarism in technological change in contrast to notions of determinism.

How ironic that at the very moment that notions of contingency and social construction of technology have triumphed among social scientists and philosophers of technology, in the world at large it appears that the experience of being swept up by unstoppable processes of technology-centered change is, in fact, stronger than it has ever been. Social scientists may call them naive, but the perception that institutions and individuals are driven by ineluctable technological change is fairly widely embraced among those who work in fields of computers and telecommunications. One of the founders of Intel, Gordon Moore, formulated Moore's Law, which states that the computing power available on a microchip doubles roughly every eighteen months. Writers on computing and society have seized upon this as the basis of their common perception that social change is now propelled by necessities that emerge from the development of new electronic technology and from nowhere else.

Writings on the emerging global economy seem similarly oblivious to the new vision of historically contingent, socially constructed, and endlessly negotiable technical options. In Lester Thurow's book *The Future of Capitalism* (1996), for example, we learn that technological change is one of the "tectonic forces" that we can only recognize and obey but not hope to master. Notions of this kind are echoed not only in the statements of businessmen and economists, but also in reports of engineering professionals at work in fields of fast-moving technical advance. In his study of the use of expert systems in industry, Todd Cherkasky notes that persons actively involved in developing and using such systems often talk in almost Ellulian terms, as if the phenomenon had a life of its own, one that transcends anyone's intentions. Even those at work on the cutting edge of research and development often talk about "where the technology is headed," suggesting that they are merely running to catch up with a process that has its own trajectory and momentum (Cherkasky, 1995).

Similarly, the literature about technology and business advises organizational restructuring and reengineering, not so much in response to technological changes upon us now, but restructuring that anticipates technological changes and acts far in advance of expected breakthroughs. In business consultant James Burrus's book *Technotrends*, there is strong advice that whatever one's focus of production is today, one must liquidate it and begin retooling in ways that incorporate new and exotic ways of achieving the same objectives. "Render your cash cow obsolete (before others do it for you)," he insists (1993, p. 353).

In writings about computers, networks, the global economy, and social institutions, there is a strong tendency to conclude that rapid changes in technology and associated developments in social practice can only be described by a reformulated evolutionary theory, a theory of biotechnical evolution. Notions of that kind inform the speculations of the Santa Fe Institute about the emergent properties of complex biological and artificial systems. Summarizing implications of this way of thinking and applying it to contemporary development in the spread of networked computing, Kevin Kelly, editor of *Wired* magazine, concludes, "We should not be surprised that life, having subjugated the bulk of inert matter on Earth, would go on to subjugate technology, and bring it also under its reign of constant evolution, perpetual novelty, and an agenda out of our control. Even without the control we must surrender, a neo-biological technology is far more

rewarding than a world of clocks, gears, and predictable simplici-
ty" (Kelly, 1994, p. 472).

In visions of this kind, one again affirms faith in the benefi-
cence of an autonomous historical process. In Kelly's view and
those of similar persuasion, the choice is neither possible, nor
desirable. In fact, attempts to impose external standards of choice
upon the internal processes of biotechnical evolution can only be
destructive.

In sum, the hope of social scientists and philosophers that arguments about social construction and contingency in technological
development would secure the domain of open deliberation and
choice is to a considerable extent contradicted by a range of experiences, perceptions, theories, and strongly advocated moral
lessons prominent among those directly involved with and excited by technological development in our time. Far from embracing the promise of humane, voluntaristic, self-conscious,
democratic, social choice-making in and around technology, a
great many observers have—for reasons they find compelling and
completely congruent with their lived experience—cast their lot
with ideas that reject or even mock choice-making of that kind.

Utopian Dreams

Descriptions of our civilization's sheer technological drivenness
are apt to strike some observers as chilling. After all, where in this
picture is there any attention to the environmental effects of a
global economy geared to limitless expansion? Where in this "out
of control" dynamism is there any care to nurture a humane civic
culture and democratic governance?

Yet many who survey the situation do not find it appalling in the
least. Yes, they may admit, the world is technologically driven, but
its trajectory leads to favorable destinations. In fact, some are
inclined to say, a new utopia is at hand.

Expectations of this kind are nothing new. Since the earliest
days of the Industrial Revolution, people have looked to the latest, most impressive technology to bring individual and collective

redemption. The specific kinds of hardware linked to these fantasies have changed over the years: steam engine, railroad, telegraph, telephone, centrally generated electrical power, radio, television, nuclear power, the Apollo program, and space stations—all have inspired transcendental visions. But the basic conceit is always the same: new technology will bring universal wealth, enhanced freedom, revitalized politics, satisfying community, and personal fulfillment. In 1856, for example, Denison Olmsted, professor of science and mathematics at Yale, wrote that science (by which he also meant what we call technology) "in its very nature, tends to promote political equality; to elevate the masses; to break down the spirit of aristocracy" (Olmsted, 1975, p. 144). Decades later, similar anticipations were inspired by the coming of the airplane. As historian Joseph Corn summarizes the "winged gospel" of aviation of the 1920s and 1930s, "Americans widely expected the airplane to foster democracy, equality, and freedom; to improve public taste and spread culture; to purge the world of war and violence; and even to give rise to a new kind of human being" (Corn, 1983).

For the past two decades this recurring dream has focused on computers and telecommunications. Again and again we hear of redemption supposed to arrive through the Computer Revolution, Information Society, Network Nation, Interactive Media, Virtual Reality, the Digital Society—the label changes just often enough for prophets to discover yet another world-transforming epoch in the works. Recently, there has been an interesting turn in this way of thinking. Familiar utopian dreams have been codified as a political ideology of sorts and given a central role in many political discussions about both American politics and world politics. What results is a pungent ideology, one that might be called cyberlibertarianism, linking ecstatic enthusiasm for electronically mediated ways of living with radical, right-wing ideas about the proper definition of freedom, social life, economics, and democracy.

This perspective can be found in a great many places. It is the coin of the realm in *Wired* magazine and other publications that key their fingers on the pulse of developments in computing and telecommunications. It can be found in countless books on cyberspace, the Internet, and interactive media, most notably George Gilder's *Microcosm* (1989) and Nicholas Negroponte's *Being Digital* (1995). Other notable writers in this strand include Alvin Toffler, Esther Dyson, Stewart Brand, John Perry Barlow, Kevin Kelly, and a host of others that some have called the digiterati. As a political program, the cyberlibertarian vision is perhaps most clearly enunciated in a publication first released by the Progress and Freedom Foundation in the summer of 1994, a manifesto entitled "Cyberspace and the American Dream: A Magna Carta for the Knowledge Age" by Esther Dyson, George Gilder, George Keyworth, and Alvin Toffler.[2] From such writings and endless on-line musings in Internet chat groups, there emerges a set of shared themes and a vision of what the new world holds in store.

First and foremost, of course, is an optimistic embrace of technological determinism, one specifically focused on the arrival of digital technologies of the late-twentieth century. A standard benchmark here is Alvin Toffler's simplistic, openly deterministic wave theory of history. Having traversed the first wave of agricultural revolution and a second wave of industrial revolution, humankind is now in the midst of third wave upheavals produced by advanced computing telecommunications. It is said to be a period in human history in which electronic information comes to dominate earlier ways of living that were based upon land, physical resources, and heavy machinery. "As it emerges, it shapes new codes of behavior that move each organism and institution— family, neighborhood, church group, company, government, nation—inexorably beyond standardization and centralization" (Dyson et al., 1994).

What conditions spawned by the new era make possible is radical individualism. Writings of cyberlibertarians revel in prospects for ecstatic self-fulfillment in cyberspace and emphasize the need for individuals to disburden themselves of encumbrances that might hinder the pursuit of rational self-interest. The experiential realm of digital devices and networked computing offers endless opportunities for achieving wealth, power, and sensual pleasure. Because inherited structures of social, political, and economic organization pose barriers to the exercise of personal power and self-realization, they simply must be removed.

Seeking intellectual grounding for this position, writers of the "Magna Carta" turn to the prophetess of unblushing egoism, Ayn Rand. Rand's defense of individual rights without responsibilities and her attack upon altruism, social welfare, and government intervention are upheld as dazzling insights by the team from the Progress and Freedom Foundation. Indeed, her portraits of heroic individuals struggling their vision and creativity against the opposition of small-minded bureaucrats and ignorant masses both foreshadow and inform the cyberlibertarian vision. Less apparent to Rand's new followers is the bleak misanthropy her writings express.

In a similar vein, the new ideology incorporates the supply-side, free-market school of economic thought reformulated by Milton Friedman and the Chicago school of economics. George Gilder, one of the authors of the new "Magna Carta," provides a crucial bridge here. His best seller, *Wealth and Poverty* (1981), helped popularize and politicize the ideas of the Chicago school during the early days of Ronald Reagan's presidency. His later book, *Microcosm* (1989), develops the social gospel of electronics, focusing upon Moore's Law as the principle that will underlie all future social change. In Gilder's view, the wedding of free market economics with the overthrow of matter by digital technology is a development that will liberate humankind because it generates unprecedented levels of wealth, a boon available to anyone with sufficient entrepreneurial initiative.

But cyberlibertarians do not argue that the wedding of digital technology and the free market will produce nothing more than a world of brass knuckled, winner-take-all competition. Instead they anticipate the rise of social and political conditions that would realize the most extravagant ideals of classical communitarian anarchism. As Nicholas Negroponte writes in *Wired,* "I do believe that being digital is positive. It can flatten organizations, globalize society, decentralize control, and help harmonize people" (Negroponte, 1995, p. 182). Just ahead is a time in which the new technology will bring sweeping structural change, fostering decentralization, diversity, and harmony. "It is clear," the "Magna Carta" exclaims, "that cyberspace will play an important role knitting together the diverse communities of tomorrow, facilitating the creation of 'electronic neighborhoods' bound together not by geography but by shared interests" (Dyson et al., 1994).

By the same token, democracy will also flourish as people use computer communication to debate issues, publicize positions, organize movements, participate in elections, and perhaps eventually vote on line. The prospect of many-to-many, interactive communication on computer networks will nurture a renewed Jeffersonian vision of citizenship and political society. When television is thoroughly linked to computing power, the universal access to cable television will finally eliminate "the gap between the knowledge-rich and knowledge-poor." In this new sociotechnical setting, the authority of centralized government and entrenched bureaucracies will simply melt away. Cyberspace democracy will "empower those closest to the decision" (Dyson et al., 1994).

Woven together from available themes and arguments from earlier varieties of social thought, the cyberlibertarian position offers a vision that many middle- and upper-class professionals find coherent and appealing. At present it seems especially attractive to white, male professionals with enough disposable income to afford a computer at home in addition to the one they use at work. It underscores many of the desires and intentions of those who see themselves on the cutting edge of technologically driven, world-historical change. What we see here are ultimately power fantasies, the power fantasies of late-twentieth-century American males, to be exact, that envision radical self-transformation and the reinvention of society in directions its devotees believe to be at once favorable and necessary.

While episodes of technological utopianism of the past have usually attracted a scant few enthusiasts, cyberlibertarianism has quickly achieved a much more prominent role. Most notable of its adherents is Newt Gingrich, leader of the "Republican revolution" and Speaker of the U.S. House of Representatives. The "Magna Carta" was the project of the Progress and Freedom Foundation, which was created by Gingrich and his followers to advance Gingrich's political program. Ideas strongly resonant with the "Magna Carta," especially those favorable to private enterprise and hostile to the regulatory role of government in the economy, occupied a prominent place in the "Contract With America," to which Republican congressional candidates pledged their fidelity during the 1994 campaign. Indeed, one of the first comments by Speaker Gingrich in the blush of enthusiasm after the successful election was a suggestion that homeless people might escape their misery if only they were given vouchers to help them buy laptop computers. In a speech to The Heritage Foundation in late 1996, the Speaker wondered, "Why can't we have expert systems and advanced computers replace 80 percent of the legal system?" and called for massive infusions of information technologies to handle much of what is now done in schools and the various fields of health care (quoted in Koprowski, 1996, p. 12).

Sometimes labeled a classic conservative, Gingrich's true position more closely resembles a radical cyberlibertarianism in which being digital and being free are one and the same. In that light, his proposals for reform in public policy strongly resemble the methods of reengineering in the corporate world, seeking to demolish structures and practices inherited from earlier times, in the hope that better ways of doing things will quickly emerge from the chaos.

Dystopian Shadows

Expecting unprecedented social benefits from the transformations they describe, today's technological utopians ignore some important questions. Who stands to gain and who to lose in the new order of things? How will power be distributed in a thoroughly digitized society? Will the institutions and practices of cyberspace eliminate existing patterns of social injustice or amplify them? Will the promised democratization benefit the whole populace or just those who own the latest electronic equipment? And who will decide these issues? About such topics, the cyberlibertarians, prophets of reengineering, and other technological optimists in our time show little if any interest. Indeed, as we have seen, some of them suspect that to ask or answer these questions could only be a hindrance to achieving the exciting next stages of "biotechnical" evolution in which human and technical life forms will merge.

Within the sketches of a world transformed by digital technology and global webs of production, however, are some distinctly dystopian possibilities. Some of these are evident in the troubling connections between work and everyday life mentioned earlier. Other ominous signs are evident within the very outlines of ostensibly hopeful visions that depict our digital future.

Celebrated in manifestos of cyberspace, for example, is the promise of a through-going dispersal of power as institutions are "demassified" in both a physical and organizational sense. But as one judges this promise, one must remember to read the fine print. Much of cyberlibertarian writing reveals a tendency to conflate the activities of freedom-seeking individuals with the operation operations of enormous, profit-seeking business firms. In the "Magna Carta," for example, concepts of rights, freedoms, access, and ownership are first justified as appropriate to individuals, but then marshaled to support the machinations of enormous transnational corporations. Crucial to its position is a concern for how to define property rights that pertain to cyberspace, a task identified as "the single most urgent and important task for government information policy." Here, the writers argue, "the key principle of ownership by the people" is the one that should "govern every deliberation." We must recognize that "Government does not own cyberspace, the people do" (Dyson et al., 1994).

One might read this as a suggestion that cyberspace is a new commons in which people have shared rights and responsibilities. But that is definitely not what the writers have in mind. For clarification they point out that "ownership by the people" simply means "private ownership." And as the discussion continues, it becomes apparent that the private entities that interest them are actually large businesses.

Thus, after praising the market competition as the pathway to a better society, the authors of the "Magna Carta" announce that some forms of competition are distinctly unwelcome. In fact, the writers fear that the government will regulate cyberspace in a way that might actually require cable companies and phone companies to compete. Needed instead, they argue, is the reduction of barriers to collaboration of already large firms, a step that will encourage the creation of a huge, commercial, interactive, multimedia network as the formerly separate kinds of communication merge. They argue that "obstructing such collaboration—in the cause of forcing a competition between the cable and phone industries—is socially elitist" (Dyson et al., 1994).

In the end, the writers of the "Magna Carta" suggest greater concentrations of power over the conduits of information because they are confident this will create an abundance of cheap, socially available bandwidth, pouring the digital solvent over what they see as hopelessly rigid, obsolete, institutional patterns. Today developments of this kind are visible in the corporate mergers that have produced a tremendous concentration of control over not only the conduits of cyberspace but the content it carries. The deregulation required by the Communications Reform Act of 1996 enables such mergers, but strong movement in that direction had begun long before the law took effect. In recent years we have seen elaborate weddings between CBS and

Westinghouse, ABC and Disney, NBC and General Electric, Turn-er Broadcasting and Time-Warner, and others. To an increasing extent, control of news, entertainment, and publishing is con-centrated in the hands of a few large concerns. What, I wonder, ever happened to the predicted collapse of large, bureaucratic structures in the era of electronic media?

Why this is problematic is suggested by the fact that during deliberations in 1995 over the telecommunications reform, CNN refused to carry advertisements critical of legislation that would allow concentration of ownership and control. In a separate inci-dent the following year, the Time-Warner corporation postponed production of a television screenplay entitled "Strange Justice," based on the U.S. Senate hearings into Supreme Court nomina-tion of Clarence Thomas, including charges of sexual harassment lodged by Anita Hill. At the time the firm had litigation before the Court challenging the "must carry" rule that requires cable television operators to carry local TV stations. Evidently, Time-Warner executive Ted Turner ordered the "Strange Justice" pro-ject shelved for fear of offending Justice Thomas and perhaps jeopardizing millions of dollars in Time-Warner profits (Schorr, 1996, p. 19).

The larger issue concerns the problems for democratic society created when a handful of organizations control all the major channels for news, entertainment, opinion, artistic expression, and the shaping of public taste. In the dewy-eyed vision cyberlib-ertarian thought, such issues are bracketed and placed out of sight. As long as we are getting rapid economic growth and increased access to broad bandwidth, all is well. To raise questions about emerging concentrations of wealth and power around the new technologies would only detract from the mood of celebra-tion.

Other points at which technological utopians distort the char-acter of sociotechnical change come in their projections about the new communities that will form in cyberspace. The "Magna Carta" looks forward to "the creation of 'electronic neighbor-

hoods' bound together not by geography but by shared interests."
Held out to readers is the promise of a rich diversity in social life.
But what will be the exact content of this diversity? The answer
soon emerges.

An important feature of life in cyberspace is that it will "allow
people to live further away from crowded or dangerous urban
areas, and expand family time." Exploring this idea, the "Magna
Carta" quotes cyberspace guru Phil Salin who argues that "Contrary to naive views, . . . cyberspaces [of the coming century] will
not all be the same, and they will not all be open to the general
public. . . . Just as access to homes, offices, churches and department stores is controlled by their owners or managers, most virtual locations will exist as distinct places of private property." A
wonderful aspect of this arrangement, in Salin's account, is that
inexpensive innovation in software can create barriers so that
"what happens in one cyberspace can be kept from affecting
other cyberspaces" (Dyson et al., 1994).

As the picture clarifies, what appears is diversity achieved
through segregation. Away from the racial and class conflicts that
afflict the cities, sheltered in a comfortable cyberniche of one's
social peers, the Third Wave society offers electronic equivalents
of the gated communities and architectural barriers that offer the
well-to-do freedom from troubles associated with urban underclass. Indeed, many proponents of the on-line world openly cele-
brate the abandonment of older cities in favor of the "wired"
exurban enclaves. For George Gilder the new promised land is to
be found in such homogeneous and untroubled locations as
Provo, Utah.

While tendencies of social separation are by no means new
(suburbanization has been with us for many decades), it is worth
noting the kinds of boundaries of occupation, residence, and
social class that define the composition of cyberspace as it currently exists. By comparison, the nexus of old-fashioned industri-
alism—the urban center—was far more diverse and socially
interactive than the on-line cultures emerging today. In the cities

it was all but inevitable that people of diverse vocations and ethnic backgrounds would have to rub shoulders with each other every day. While the very wealthy were able to shelter themselves in mansions in remote locations, the rest of the populace was forced to contend with social differences on a daily basis. A mirror of these encounters was present in the general interest newspapers that served as a primary means of communication. While people at different levels of society read papers with drastically different slants—from sensationalistic tabloids to serious, high-quality journalism—the report was always about the same social universe: the metropolis situated in the wider world. How different this is from the smug, self-contained yuppie cyberzines—*Hot Wired, Slate, Salon,* and others that fill pages on the World Wide Web.

There are, in my view, signs that on-line benefits of access to information and on-line community are being purchased with a decline in habits of sociability. Because we are citizens of cyberspace, even our next door neighbors do not matter all that much. We can stay in our rooms, stare at flat screens, surf the Internet, and be satisfied with simulacra of human contact. Recent reports indicate that this mentality has already affected social life on college campuses. Rather than congregate in coffee houses or other gathering places, many students stay in their rooms or in computer labs communicating through the network, even if the other persons in the conversation are no more than an arms length away (Gabriel, 1996). At the college where I teach, it is not uncommon to find young women and men who are far more comfortable with the disembodied relationships in the global cybersphere than they are with persons who are physically present. Thus, all-too-often becoming "wired" involves increasing isolation, discomfort, and even fear of the presence of other people.

My fear is not that people will forget what love is about or reject the pleasures of human company. I have more faith in biology than that. What worries me is that people will begin to employ networked computing as they already use television, as a way to "stay in touch" while avoiding direct contacts in the public world. The basic question concerns how we will regard ourselves and others in a wide range of technically mediated settings. Will people beyond our immediate family, professional colleagues, and circle of on-line friends be seen as connected to us in important, potentially fulfilling ways? Or will they be seen as mere annoyances, an unwanted human surplus that needs to be walled off, controlled, and ignored?

These questions are especially important when it comes to those in the United States who are already seen as candidates for the discard pile—the poor, disabled, and working-class elderly, among others. For a significant percentage of young black males, for example, the digital electronics most likely to affect their lives are the sophisticated surveillance mechanisms built into today's "control unit" megaprisons, the infrastructure of an American gulag. What does the emerging utopia of cyberspace and global production hold for them?

On occasion, even the most avid proponents of rapid technological and organizational restructuring pause to reflect on those left high and dry by these transformations. At the conclusion of his book *Beyond Reengineering*, Michael Hammer worries whether the program he proposes will bring "utopia" or "apocalypse." "What will become of the people who merely want to come to work, turn off their brains, and do what they're told until quitting time? Of those who simply don't have the drive, ambition, and intensity to focus on processes and customers? . . . What of those who can't handle constant change, who need stability and predictability? Must they all be left behind, orphans of the new age?" Hammer ponders the prospects for education and retraining to raise obsolete workers to the levels of ability and initiative that will be required of them. But he laments that such improvement seems doubtful even over the long term. "The problem of what to do with 'little people' will be with us for some time" (Hammer, 1996, pp. 259–60). It is a major problem indeed.

Direct Engagement with Technical Things

Episodes of social upheaval linked to technological change have been with us for a long while. During the past two centuries there have been a number of ways people have responded to vexing disruptions in their ways of living—labor union organizing, Luddism, populism, socialist politics, issue-centered movements for social reform, environmental protest, and Green politics among others. While these approaches still have much to offer, none of them seems fully prepared to confront the challenges presented by the powerful, polymorphic, destabilizing forces contained in technological innovation today. Protest of past decades were often able to focus on relatively fixed targets—obnoxious railroads, industrial assembly lines, controversial water systems, toxic waste dumps, and the like. Many intellectual critiques of technology, similarly, lamented the ponderous rigidity of technology-centered institutions—Max Weber's Iron Cage of bureaucratic rationality or Lewis Mumford's lead-footed Megamachine, for example. But today intellectual and political strategies must recognize the sheer transience of instrumental and organizational forms as well as the plans that guide them. In this respect Marx was entirely prescient: "All that is solid melts into air."

The condition we face has strong implications for thought and action. Those who care about human well-being and the values of civic culture must be prepared to confront emerging technologies directly, early on in their development. To an increasing extent the crucial questions about the complexion of work, education, leisure, and community life must be engaged far "upstream" in processes of sociotechnical planning, design, and development. No longer will it suffice to seem ignorant or surprised as the new technical devices are woven into the social settings one cares about—computers in schools, agile technologies in the workplace, Web browsers in the living room, or surveillance cameras in the mall. Instead one must focus upon important areas of shared purpose where new devices might intervene and become involved in processes of change.

To my way of thinking a perfectly valid way to become involved is simply to say "no." For there is nothing more positive than to resist technically embodied schemes predicated solely on efficiency, productivity, profit, or the dubious promise of some desirable effect (for example, better schools) while ignoring the deeper virtues already present in structures and practices scheduled for hasty renovation or elimination. Positive strategies of this kind are present, for example, in the many cases of local resistance to the coming of Wal-Mart megastores. This may not seem like a case of intervention in the preparation of a technological system, but actually it is. Among the devices most influential in our economy are sophisticated electronic data systems that enable instantaneous inventory control. Wal-Mart and similar chain stores are based upon a seemingly innocuous digital spinal cord that enables the chain to have precise knowledge over the flow of goods that enter and leave its outlets. This has the consequence of reducing the funds invested in inventory at any given time. When combined with the advantages that accrue to large, multi-unit retailing, such systems enable the Wal-Marts of the world to undercut small, local retailers. Of course, people look at the falling prices. Joe's Downtown Pharmacy sells the toothpaste for $3.00 and Wal-Mart for $2.30. What's to choose? Let's buy the lower cost item, provided by the more efficient seller.

But as increasing numbers of people across America have begun to notice, the cost of the tube of toothpaste is not really $2.30. The costs must also be measured in broader social consequences. Small local retailers provide a key link in networks of social support and webs of civic vitality. When the large chains move in and the small businesses die, not only are many jobs lost, but also communities that housed them begin to wither. When compounded by other forces that tend to weaken crucial supports for community life, the effects of electronic data systems contribute to the growing sense that the places where we live are no longer friendly, safe, or humanly sustaining. That is why saying "no" to otherwise appealing "developments" like the building of chain stores and the spread of shopping on the Internet makes perfect sense in some cases (see Winner, 1997).

Being involved with upstream choices, however, often means becoming knowledgeable about the design of new systems in the hope of shaping their features. After decades in which labor unions ceded control of almost all technology planning to corporate managers, some unions have decided to cultivate new expertise about which production systems are in the works and to play a role in their design. Exploring ways to influence the hardware, software, and social arrangements of new workplace technologies comes under the general theme of "high performance systems." As a report by the Work and Technology Institute in Washington, D.C. comments, "The core element of high performance is to give front-line workers the responsibility, autonomy, and discretion for key decisions at all points of production and to provide employees with the information, skills, and incentives needed to successfully exercise those judgments" (Jarboe and Yudken, 1996, p. 2). Whether or not these efforts flourish in the troubled waters of American labor relations remains to be seen. But the decision to engage the shape of production technologies directly, rather than let them flow as if from a volcano, is an important turn in labor's understanding of its horizons.

There are other cases that might be cited as lively examples of upstream engagement with technological change—the creation of civic networks in computing, the development on implements tailored to the needs of the disabled, the rise of community supported biodynamic farming, and the firestorm of protests that have greeted attempts to assemble and market databases with stored information on millions of consumers. While these are only small bubbles within much larger tides, they do reflect some willingness to address issues about the common good at junctures where new devices, techniques, and systems are in the making.

Of course, a renewed awareness and willingness to act will not be enough. Occasions for participation in technology-shaping must be discovered, created, or forcefully demanded. In most cases, the origins and character of impending technological change are opaque to workers, consumers, and ordinary citizens because they have never been included in research projects, engineering designs, or business plans, including ones destined to alter their lives profoundly. This is certainly true of those who click their keyboards day after day at the Remote Encoding Center in Latham. No one asked for their ideas on what the facility and its equipment might look like or how it would (or should) affect them. Their experience, like our own, is not that of having rich opportunities for study, experiment, and choice in the "social construction" of technology. Instead it is the experience of imposed solutions, of being receptacles for patterns and processes whose character has been decided elsewhere.

Notes

[1] See, for example, Wiebe Bijker et al., 1987. For my critique of this way of thinking, see "Social Constructivism: Opening the Black Box and Finding It Empty" (1993).

[2] Esther Dyson, George Gilder, George Keyworth, Alvin Toffler, "Cyberspace and the American Dream: A Magna Carta for the Knowledge Age," Release 1.2, Progress and Freedom Foundation, Washington, D.C., August 22, 1994, at http://www.townhall.com/pff/position.html. This document was published to the World Wide Web where there is no standard style for pagination. All references that follow will simply indicate that a quote is somewhere in this "Magna Carta."

References

Bijker, Wiebe, et al., eds., *The Social Construction of Technological Systems: New Directions in the Sociology and History of Technology* (Cambridge, Mass.: MIT Press, 1987).

Brand, Stewart, "Two Questions," in "Scenarios: The Future of the Future," *Wired* (December 1995).

Burrus, Daniel, *Technotrends: How to Use Technology to Go Beyond Your Competition* (New York: HarperBusiness, 1993).

Casells, Manuel, *The Rise of the Network Society* (Cambridge, Mass.: Blackwell Publishers, 1996).

Cherkasky, Todd D., "Obscuring the Human Costs of Expert Systems," *IEEE Technology and Society Magazine* 14 (Spring 1995): 10–20.

Corn, Joseph J., *The Winged Gospel: America's Romance with Aviation, 1900–1950* (New York: Oxford University Press, 1983).

Dyson, Esther, Gilder, George, Keyworth, George, and Toffler, Alvin, "Cyberspace and the American Dream: A Magna Carta for the Knowledge Age," Release 1.2, Progress and Freedom Foundation, Washington, D.C., August 22, 1994, at http://www.townhall.com/pff/position.html.

Ellul, Jacques, *The Technological Society,* John Wilkinson, trans. (New York: Vintage Books, 1964).

Gabriel, Trip, "Computers Help Unite Campuses but Also Drive Some Students Apart," *New York Times* (November 11, 1996).

Gilder, George F., *Wealth and Poverty* (New York: Basic Books, 1981).

Gilder, George F., *Microcosm: The Quantum Revolution in Economics and Technology* (New York: Simon and Schuster, 1989).

Hammer, Michael, and Champy, James, *Reengineering the Corporation: A Manifesto for Business Revolution* (New York: HarperBusiness, 1993).

Hammer, Michael, *Beyond Reengineering: How the Process-centered Organization is Changing Our Work and Lives* (New York: HarperBusiness, 1996).

Jarboe, Kenan Patrick, and Yudken, Joel, "Smart Workers, Smart Machines: A Technology Policy for the 21st Century," Work and Technology Institute, Washington, D.C., 1996, p. 2.

Kelly, Kevin, *Out of Control: The New Biology of Machines, Social Systems and the Economic World* (Reading, Mass.: Addison-Wesley, 1994).

Koprowski, Gene, "Gingrich Proposal: Lets Delete All Lawyers . . . ," *New Technology Week* (December 9, 1996): 12.

Lohr, Steve, "Though Upbeat on the Economy, People Still Fear for Their Jobs," *New York Times* (December 29, 1996).

Marcuse, Herbert, *One-Dimensional Man: Studies in the Ideology of Advanced Industrial Society* (Boston: Beacon Press, 1964).

Mumford, Lewis, *The Myth of the Machine: The Pentagon of Power* (New York: Harcourt Brace, 1970).

Negroponte, Nicholas, "Being Digital, A Book (P)Review," *Wired* (February 1995).

Negroponte, Nicholas, *Being Digital* (New York: Alfred A. Knopf, 1995).

Olmsted, Denison, "On the Democratic Tendencies of Science," in Thomas Parke Hughes, ed., *Changing Attitudes Toward American Technology* (New York: Harper & Row, 1975).

Rostow, W. W., *The Stages of Economic Growth: A Non-Communist Manifesto* (Cambridge, England: Cambridge University Press, 1960).

Schorr, Daniel, "Strange Justice," in "An Age of Media Mergers," *The Christian Science Monitor* (October 22, 1996).

Thurow, Lester C., *The Future of Capitalism* (New York: William Morrow and Company, 1996).

Uchitelle, Louis, and Kleinfield, N.R., et al., "The Downsizing of America," a seven part series, *The New York Times* (March 3 through March 9, 1997).

Warme, Barbara, et al., eds., *Working Part-time: Risks and Opportunities* (New York: Praeger, 1992).

Winner, Langdon, "Who Will We Be in Cyberspace?" *The Information Society* 12:1 (January/March 1996a): 63–72.

Winner, Langdon, "From Octopus to Polymorph: the Moral Dimensions of Large Scale Systems," in Lars Ingelstam, ed., *Complex Technical Systems* (Stockholm: Swedish Council for Planning and Coordination of Research, 1996b), pp. 183–96.

Winner, Langdon, "Social Constructivism: Opening the Black Box and Finding It Empty," *Science as Culture* 3:16 (Fall 1993): 427–52.

Winner, Langdon, "The Neverhood of Internet Commerce," *Technology Review* (August/September 1997).

PART 2

Keynote Address:
Technology and Culture

Technology and the Rest of Culture: Keynote

BY ARNO PENZIAS

I have trouble understanding the title to my own article. What do these two much-used terms—*technology* and *culture*—really mean? My favorite definition of *technology* comes from John Kenneth Galbraith: the application of organized knowledge. So far so good. But what about *culture*? Based upon my 1950's education, culture is whatever Margaret Mead, or perhaps Ruth Benedict, said it was. Something like what your parents make you believe before you're old enough to form an independent opinion.

Let me begin by focusing on the first of these two terms and seeing what it reveals with respect to the second. Among all the stuff that human ingenuity has produced, information technology has probably had the greatest impact upon culture. Just consider how the invention of, say, vowels—which is indeed a technology—shaped Greek civilization (and also the good press that they have gotten afterwards, as a result of it). The technology of today abounds in similar examples. Today's world is so pervaded by information technology that we're beginning to fuse the two together in our minds. Perhaps we should talk about "cyberspace" to avoid confusion with other stuff such as locomotives, hydroelectric dams, and dental drills. But as far as we're concerned, information technology so pervades our world today that we lose little by restricting ourselves to that area. So let me restrict myself to the information aspects of technology in trying to answer, or at least address, a few simple questions concerning

79

technology and culture: Where are we? How did we get here? What's out? What's in? And, finally, What lies ahead?

Where Are We?

In looking at today's cyberspace technology, three important elements emerge. The first is the *overwhelming use of digital information storage*. Most of the records of our civilization today are created on computers—either as text or, increasingly, in image form as well. While the new digital cameras currently seem an expensive novelty, pretty soon carting out film to some little place in a drugstore to get your pictures developed will be a thing of the past. But digital photography offers far more than merely avoiding an errand or two. With digital cameras, it's never too late to change all those almost-good snapshots in which someone hadn't remembered to button a coat or take off a name tag. In a digital picture, anyone can just sit down at a computer and give the subject a sartorial makeover. While that novelty will wear off over time, it will initially sell a lot of cameras. More important, digital technology will offer consumers and professionals with an inexhaustible supply of "film," together with truly prodigious quantities of convenient storage space.

While the digital creation and storage of images is growing rapidly, text-based material still fills the bulk of today's storage space—comprised largely of records of computer-supported transactions. The sheer size of this ever-expanding mass of data staggers the imagination. Every day, more and more material produced and stored in computer-accessible, readable, and modifiable form. Moreover, the older material—the paper—that exists in our world's archives is increasingly being scanned into its digital equivalents. And so it is that digital information of all kinds, in enormous amounts, is becoming available to its owners—and those they choose to share it with.

An article on mass data storage in a recent issue of *Computer World* speaks in terms of terabytes, or trillions of bytes. For text-based data, each byte equates to an alphanumeric character, or keystroke. So a terabyte, or trillion keystrokes, represents some four hundred million pages of text. Today, many businesses must cope with a terabyte of data, with the largest companies facing ten times that much. Where will it all end? Who knows? Tales of future hundred-terabyte systems pepper the table talk in Silicon Valley cafes.

Alongside digital information storage, there is *the growing interconnection of the world's computers*. Computers have played a series of roles, beginning with numerical calculation. The first programmable string manipulation machines—or computers—earned this catchier title by serving as replacements for real computers. Originally, *computer* was a job title given to people who sat at desks, read numbers written on a pad, punched them into mechanical calculators, and wrote down the results. So if you had happened to visit Los Alamos during World War II and happened to ask "Where are the computers?" your host might have replied "They're just coming back from their lunch breaks," or "They're off today because it's a holiday." Sounds strange, doesn't it? While other machines have taken on human tasks—the term dishwasher comes to mind—none other so completely as to make the original usage sound absurd.

And so, a machine that could push its own buttons and remember what it had done began its career by crunching numbers. While retaining its original name, the computer has taken on additional tasks, such as "data processing." This new role began when memory got cheap enough to allow the introduction and processing of text by such machines.

In that spirit, today's computers have taken on a new role, acting as windows on the world for their users. So we have the advent of today's networked world. Not just of the Internet, but *intranets* within corporations, *extranets* that link selected intranets, plus all the public and private voice and data networks that serve the general public, corporations, government agencies, and not-for-profits. It seems little short of miraculous that tens of millions of these computers can exchange information with one another on a reasonably dependable basis.

The third element is the *emerging dominance of electronic information sources*. There was a time when people got their information from other folks. Spoken and written words were dominant, and people relied on newspapers, magazines, archives filled with paper correspondence, libraries, stories, and even rumors. But so much is now predicated on the explosive growth of information appliances. Next year, according to some pundits, the world will produce one hundred million personal computers. A hundred million personal computers. One sixteen-megabit memory chip in one of those computers contains at least that number of transistors—each sort of equivalent to a vacuum tube. The memory board of each machine probably holds a dozen such chips, or about two hundred million transistors. When we start thinking about all the other devices in that one desktop computer, our total might climb to something like a billion transistors.

Winston Churchill called World War II the "Wizard War" because of its technology—radar, sonar, smart torpedoes, and a whole bunch of other electronic innovations. World War II was fought with something like a billion electronic devices per side; today, the number of transistors in a single personal computer exceeds the number of vacuum tubes employed by the entire United States during all the years of that global struggle.

As information appliances grow in number, their use grows even faster. The number of computers produced annually is going up by something over 20 percent a year, doubling, therefore, every two to three years. But the Internet doubles the amount of information it ships from one place to another every few months, and futuristic projections in recent magazines about the Internet always mention the same underlying algorithm: everything will at least double every single year for as long as the writer cares to predict.

Alongside the technological features of this landscape, there are cultural ones as well. What can be said about *cyberspace culture?*

Having recently moved to California's Bay Area, I feel sort of like Margaret Mead. I've moved out to San Francisco to study, and to take part in, the workings of Silicon Valley. The definition of culture I use to study Silicon Valley lists four attributes: a culture 1) evolves among a distinct people; 2) is received through inheritance or upbringing; 3) makes moral demands; and 4) produces art.

How do these attributes apply to Silicon Valley? As the home of a distinct people, it's not an accident that this uniquely powerful incubator of technology has appeared in the least-rooted region of the world's least-rooted country—the place where most people have just come from somewhere else. Among countries, the United States seems least tied to tradition, and within our country, where but in California do iconoclasts abound in such numbers?

Is this culture received through inheritance, or upbringing? They do talk about generations out there, but they're not the traditional ones. Cyberspace's generations outpace biology by a wide margin. People out there think in terms of Moore's Law: every year and a half, the number of components on a chip doubles, and a new product generation emerges.

Does the culture make moral demands? Yes, as evidenced by the limits on vaporware. *Vaporware* is a Silicon Valley term for products that are in the pipeline but have yet to be developed. Vaporware allows companies to express their aspirations, rather than just hard-and-fast accomplishments. Properly used, announcing vaporware can keep customers and investors happy, and discourage would-be competitors from venturing onto one's turf.

Silicon Valley executives seem surprised at how indignant the folks in other parts of the country get when they take one of these claims too seriously. On the other hand, a company recently made news by actually producing a product on the same day that they had announced it. Apparently, they had pushed the community's tolerance a little bit too far in the past—so they had to make amends to local moral demands.

Finally, concerning the production of art, quite seriously, I think that cyberspace contains a form of art in the construction of elegant metaphors—somewhat like collage, the creation of beautiful constructs from seemingly mundane constituents. The World Wide Web, for example, was created by an inspired human who, looking at an empty space, filled it with a series of logical operations, putting them together in such a way that others say, "Wow, that's beautiful" and—more important—in a way that causes others to say, "Why didn't I see that?"

This is what happens when we go to a museum: people look at great art and say, "Why didn't I see that?" I had that experience when I looked at Van Gogh's *The Road Builders*. While trying to make some sketches of that celebrated masterpiece, it suddenly dawned on me that the gnarled muscles of those men bent over their shovels were echoed in the bent trees lining the road on which they were working. And so I caught a glimpse of the artist's genius in seeing and capturing that relationship. Seeing something others don't see. In that spirit, I think it's fair to say that elegant constructs represent art in a different kind of medium. Allowing for its idiosyncracies therefore, we can perceive the attributes of culture.

How Did We Get Here?

The current state of affairs was achieved by a march of technology unprecedented in it's size and scope, beginning with an *exponential growth of microprocessor power*. This process can be illustrated by the story of the scholar who asked a king for chessboard full of wheat, with a single grain of wheat on the first square, two on the second, four on the third, and so on. By the end, having doubled the number of gains just sixty-four times, the chessboard would contain some ten-billion-billion grains. More than the kingdom's gross national product in value. Nothing can double indefinitely, of course, but exponential growth can produce quite a ride while it lasts.

The realm of possibilities resulting from this growth can be hinted at by the fact that the actor who played Forrest Gump's friend really had legs. The illusion was achieved with a computer, pixel by pixel (pixels are microscopic picture elements). Think of each frame of digitized film as a mosaic of infinitely small tiles, with each picture element so small that the mosaic looks as if it were painted with continuous brush strokes—or a photograph of a real object. In this instance, each of the "tiles" corresponding to the supposedly legless actor was subtracted out of the picture, and replaced by another tile from a picture of the identical scene with that actor missing.

The number of mathematical operations required is astronomical. Identify a location, go to the corresponding memory location, replace the value stored there with one from another table. This sounds simple—if we were discussing a color-by-numbers picture, we could probably do it by hand. But think in terms of millions of picture elements per frame and hundreds of thousands of frames per film. Operations quickly mount into the trillions for that one task alone, the achievement of which is only possible because of this exponential growth of processor power.

Another factor in today's computing environment is the *plummeting cost of computer memory*. When I first mentioned that "cheap" computer memory changed the world of computing by enabling data processing, I had the dollar-per-bit memory of the IBM 704 in mind. When that machine came on the market in the 1960's, it offered 128,000 eight-bit bytes for a mere million dollars. In those days, a million dollars would buy you an apartment house just off Central Park West. Today, on the other hand, the purchase price of a million bits worth of memory would only suffice to make the smallest of repairs to such a building. Today, if you went to a hardware store with the cash equivalent of a million bits worth of memory, you'd probably have to get the clerk to open a blister pack and give you a replacement for one of the building's wood screws; something as expensive as a doorstop would be totally beyond your means. Imagine the price of an entire apartment house collapsing to that of a single flat-head brass wood screw.

I once used this analogy on Gordon Moore, author of Moore's Law, but he topped me without batting an eye. "I remember when memory was a dollar a bit *per month*," he told me, "because we had to replace the vacuum tubes." Think about what that would mean in terms of cost. A modern, twenty-megabyte personal computer contains over one hundred million bits in its memory. Imagine running such a personal computer with 1950s' technology: two billion dollars per year for memory maintenance alone, and not in 1997 dollars, but in much bigger 1955 dollars. That illustrates the pace of today's continuing decline in the cost of memory.

The use of computers has also been facilitated by *global deployment of optical fibers*. Our world has become girdled by strands of optical fibers, each one capable of carrying tens, and soon hundreds, of billions of bits per second almost anywhere. What does a hundred billion bits per second mean in practical terms? Enough transmission capabilities to carry one million simultaneous telephone conversations—each one a private conversation with someone on the other side of the world. While the cost of installing an optical fiber cable can get pretty expensive—especially when deploying one across an ocean—the fiber strands themselves have become relatively inexpensive. So inexpensive, in fact, that a length of this ultratransport glass doesn't cost a lot more than some kinds of kite string in an upscale store. So now these multibillion bits form a networking fabric—the lanes, if you will, of the much-discussed Information Superhighway.

Finally, computing ability has been enhanced by the *growing power of modular software*. Consider the challenge of producing bug-free programs. Such work is tough enough when it encompasses as much text as an essay, or a high school textbook. But programs frequently contain millions of lines of code. How can one possibly test all possible interactions in advance?

Fortunately, most of the software we encounter behaves itself rather well because programs are increasingly modular; that is, they are constructed from "chunks" of other, previously tested programs. When you send out a query on the Internet, for example, you launch a program that sifts through electronic nooks and crannies around the world. Usually the only problems stem from the user not being specific enough about what is wanted. The program itself has little trouble making itself understood by the computers it encounters. When I recently wanted to check on a Bell Labs product called "Inferno," for instance, I got back pointers to over 20,000 published references to Dante's poem, in addition to the material I wanted. I also got back a couple of so-called "applets"—small programs capable of running on my machine—offered by information providers whose companies also happened to be listed under that same title. And all from one little query message.

The interesting thing here was that I had no advance knowledge of the various kinds of computers my query would reach, who owned them, or who programmed them. Moreover, the folks who wrote the applets, and organized the data I received in return, knew equally little about my computer. Yet the material I asked for arrived in usable form, with no problems with compatibility or unforeseen interaction. The modularity of the software involved reduced the problem to one of swapping building blocks that vary in their internal makeup, but plug into one another like the pieces of a *Lego* set.

How Did We Get Here?

There's more to the technology story than the growth of its parts. Today's spectacular progress in individual technologies merely represents the tip of the iceberg. Throughout modern history, the most significant impacts of technology have come from its combinations, and today the dramatic changes taking place largely stem from *technology mergers*.

One example is what happened to three familiar appliances once the addition of microprocessors gave them the power to negotiate with one another: facsimile machines, modems, and radio telephones. Each one of these devices languished for more than a generation in its original form. Facsimile machines have been around since the 1930's; modems appeared shortly after World War II; and radio telephones are well over sixty years old, but the use of these three devices has only taken off in the past ten years.

Today, people wonder how anyone could have conducted business without a fax machine, but for about half a century the technology was used by only a handful of people. What happened? Why did these things only take off now when they had been out there for so many years? What changed? The basic fax machine continues to do its job. But, in former times, potential users faced a world filled with a whole bunch of different facsimile machine models. Each model could only "speak to" one of its siblings. Moreover, uncertainties in the quality of telephone lines sharply restricted transmission speeds. Today, on the other hand, each one starts by chirping the message: "I'm a fax machine," and listens for "That's good. I'm a fax machine too, and I can deal with the following formats." "How about Group IV?" "Fine with me." "Let's do it!" "I can transfer data at the following baud rates." "I can go only up to a baud rate of x." "Beep. Beep. Beep. Chiiiiiirp." Why bother translating? Just listen to it next time, and ponder the fact that those two machines are negotiating with one another, so as to transfer images most efficiently.

And that same underlying technology allows computer modems to interoperate at high speeds. The first thing that a high-speed modem does when it gets on a telephone line is tell that line: "Turn off your echo canceler. Don't help me." Because modern telephone lines employ a special circuit that prevents the annoying echoes of former days. (Years ago, folks calling Europe, for example, generally heard an echo of their spoken words. Today, special electronic devices kill such echoes before you can hear them. They make the system sound echo-free by listening for echoes and canceling them electronically.) The last things the modem wants is that kind of "help," so the first thing any modem must do is turn that canceler off. Then it starts trading information with the modem on the other end. It sounds different from a pair of fax machines starting up, but the principle remains the same. My first modem had a two-position switch, for low speed and high speed (high speed was 1.2 kilobits). But today, modems negotiate, working at the best data rate that the line can transmit.

Radio telephones, which are now called cellular phones, have benefitted from the same technology. In the old days, a radio telephone would barely fit in the trunk of a car, let alone in someone's pocket. But again, because new phones negotiate with local cell sites, together they can decide which local radio beacon the phone ought to use, and at which frequency; they authenticate the user's right to service and, with the newest phones, even decide how much power to put out. These days, phones try not to drown out the neighbors, so that they all may use the spectrum as efficiently as possible.

Another technology merger underlies the advent of the World Wide Web, which stems from the merger of at least three technologies. Before the Web, the Internet was used by expert techies for things such as electronic mail, the transfer of data files, and an application called *telnet,* which allowed sophisticated computer users to connect to distant computers.

The Internet provided a fabric that was primarily good for techies and a few others, but then, new software, together with fast modems and high-resolution personal computer screens that could handle nice pictures, changed it dramatically. When those three technologies came together, we got the World Wide Web. All of a sudden, ordinary people could access the Web, which offered attractive graphics in place of dreary rows of typed text. In such fashion, the merger of computing and communication has created this new environment we call "cyberspace."

What's Out?

The advent of cyberspace places us in the middle of a revolution: out with the old and in with the new. A recent cartoon in the

New Yorker pictured today's pace of change in terms of a well-dressed couple riding in a taxi cab with anxious looks on their faces. The caption said: "Step on it driver, this restaurant may be out before we get there." "Out" presumably meant, "consigned to the out list" or "out of style."

Today's "outs" include *industrial economies of scale* and *hierarchical organizations,* together with their *investments in mechanical technology.* Picture a sequence of rigidly controlled machines, with each stage in the sequence assigned some task repeated over and over again. In the case of automobile manufacture, say, you might start with a bare chassis, drill holes for engine bolts, add the mounts, the engine block, a cylinder head, and so forth. As long as every station along the way does its task with rote reliability, the production line works well. No room for individual initiative. Everything controlled from the top. In bygone days, such hierarchical investments in rigid mechanical technology prompted Henry Ford to say, "My customers can have any color car they want, as long as it's black." Imagine such a philosophy in today's market.

Out also is *organizationally generated paperwork.* Say you worked in a university, a business, the government, or whatever. If you needed something really quickly, and you were lucky enough to find a helpful human being, that person might well have said, "Okay. Here. We'll take care of it and now, fix the paperwork later." Getting the task accomplished seemed the easy part. Paperwork frequently took on a life of its own.

But organizationally generated paperwork is on it's way out. Think of all the paperwork generated by dealing with a bank teller. Slips to be filled out, checked, stacked, rechecked, moved, sorted, read, and so on. All that back-office work disappears in a world in which customers can just punch buttons on an ATM. In that way, customers can effectively access the bank's ledger directly, and draw cash right out of the vault. All that internal paperwork has disappeared.

Another "out," one whose disappearance will, I think, have profound importance, is *erosion of dedicated resources.* In the bazaars of

Byzantium, each purchase would trigger a long negotiation. I remember going to the Middle East in the 1960s and visiting a Bedouin encampment on market day. I saw an old coin I wanted to buy and learned that its owner wanted the equivalent of seventy-five U.S. cents—about three times the amount he actually expected to get for it. Before I knew it, I was sitting opposite him, with a referee at our side to help the bargaining along. I'm afraid I didn't do much for peace and understanding in the region. I ruined the merchant's day by pulling the equivalent of seventy-five cents out of my pocket and giving it to him right away. So much for a time-honored way of conducting business.

This misunderstanding can be blamed on my Western notion of efficiency: all I wanted to do is get up from the carpet, get my coin, move on to the next display. Unaccustomed to the overhead exacted by prolonged negotiations, I wanted to conduct business in the manner I'd grown up with. Ever since Adam Smith pondered the wealth of nations at the dawn of the Industrial Revolution, division of labor has outperformed individual producers. Traditionally, the value of each employee's contribution has been fixed in advance and exchanged freely within the bounds of the enterprise.

In practical terms, corporations might not have had the world's most efficient library, secretary, or stockroom, but, because these services came from dedicated resources that everybody in the company used, the employer saved the staff time that would have been consumed by per-transaction negotiations. This arrangement made a lot of sense from an employee perspective as well. The boss wouldn't fire someone who got sick for a couple of days, for instance, because the cost of hiring a replacement presented too much of a barrier. Imagine workers having to submit their skills to an auction every time the need for some task sprang up.

Now, however, that exclusive reliance on dedicated resources no longer goes unquestioned. Bargaining, and the other elements of transactions, have all become very, very cheap. Most corporate libraries now charge users for their services on a per-item basis, for instance, and occasionally lose business to outside suppliers.

As transaction costs plummet, *mass media's economies of production and distribution* move toward the "out" column, as do *distance-related barriers to commerce.*

What's In?

What's in? Today's economies benefit less and less from scale, and more and more from *economies of interoperation.* For example, an electronic gadget called Palm Pilot, a pocket-sized electronic notepad, will hold some 2500 addresses, a five-year calendar, and handy things like a "to do" list. While it's quite useful, the important point at issue here is its economies of interoperation: primarily, the fact that it costs the manufacturer less to produce this gadget (because it comes with a means of connecting to a personal computer) than it would to produce its stand-alone equivalent. Whenever the user needs to create an up-to-date backup copy of the data stored in the Pilot, or add new information from another source, it can be placed in a little cradle attached to a personal computer and files then can be transferred back and forth with ease. This interoperation with a personal computer makes the hand-held unit simpler and cheaper. Instead of including a rechargeable power source, for example, it makes do with a couple of inexpensive triple-A batteries. It doesn't have a keypad, nor on-board backups or elaborate technology for error recovery. All those functions are located elsewhere, on the personal computer.

As a result, the interoperation benefits the manufacturer as well as the user, which makes a big difference. In a competitive market, the economies of interconnection will incite more and more producers to interconnect their offerings. Since it will be so much cheaper to create interconnected products, consumers will have to pay extra to get stand-alone ones, resulting in a world in which everything is connected to everything else.

What else is in? *Direct information access:* the world-wide links to people, data, and machines. Today, there is also an increase in *outsourcing via low-cost transactions.* Take Wal-Mart, for instance. They have outsourced their purchasing, and no longer buy anything. Wal-Mart no longer needs a conventional purchasing organization anymore, because their suppliers handle most of that operation for them. Every time a box of Pampers produces a "beep" on the check-out counter's bar-code reader, for example, Wal-Mart credits Proctor and Gamble's account with the wholesale price of that package. Transactions have become so cheap that they can be handled independently.

Micromarketing via massive databases is also in. While many people believe that every time someone tries to create a large database of consumer information, there's an uproar, that's not true. Massive consumer databases are continually sprouting all over the place, some of which are even available free on the Internet. More important, huge amounts of data about (almost) everyone and everything are available for sale everywhere. Mining that data allows for micromarketing—knowing enough about an individual to custom-tailor an offer to that person or small group of people.

But small isn't always beautiful. Micromarketing also allows zealots of all kinds to identify, recruit, and organize small groups of like-minded people. In the past, someone might have been ashamed of being the only person in town who harbored some particular kind of bigotry. Whatever hatred one might harbor today, however, can find positive reinforcement in its own (virtual) community, thanks to communication technology.

And finally, instead of the geographic barriers of the past, we have *global products for global consumers.* Of all the places in the world, I can think of none with a more durable culture than the one we see in France. Yet when Disney produced its version of *The Hunchback of Notre Dame* recently, they had a special display at the Victor Hugo Museum; and the French loved it. Instead of Charles Lawton, they flocked to view mementos of this happy-go-lucky guy who sings with the birds.

What Lies Ahead?

Even though global communications have shrunk distances, there remain *continuing advantages for physical communities.* For example, Silicon Valley's venture capitalists can locate anywhere they want. They have plenty of money, so would-be investees ought to be willing to come to them. They have video conferencing, and all kinds of other telecommuting stuff. And yet, one of the largest of the venture capitalists—a very successful company— recently moved its office some forty miles, from downtown San Francisco to Palo Alto, thereby adding that distance to the daily commute of the company's principal owner and many of the other partners. Why? They felt that San Francisco was "too far away from the action," so they had to move their offices.

The action referred to takes place along Sand Hill Road, a kilometer-long stretch of highway, just off Interstate 280. A single cluster of office buildings contains most of California's significant venture capitalists. While they compete fiercely with one another, they also cooperate with, and learn from, one another as well. Physical proximity builds trust, and trust leads to mutual benefits.

Communities of this kind pop up in quite unexpected places. Even though China boasts the world's oldest and largest ceramics industry, for example, its luxury hotels use imported tiles in their bathrooms. "Tile is tile" you might think, but there's enough of a difference to cause hard-headed investors to cart this particular item halfway around the world because the center of high-quality tile happens to be concentrated in a small region around Bologna, Italy. Anybody who wants to be really good at manufacturing high-quality ceramics locates there—it's the Silicon Valley of tile.

In both these cases, community members cluster in order to benefit from physical proximity, but what does this provide? Among other things, neighbors provide one another with means of nongovernmental recourse. It's not just a matter of "sue me if you don't like it"—you have to face your neighbors if you act badly. In Silicon Valley terms, your physical presence shows you're more committed, willing to show what person you are through your day-to-day behavior. In the Bologna case, your children may well wish to marry someone from that community, so you'd better not be a bad actor. Neighbors feel freer to share information, as well as infrastructure, so learning and sharing takes place.

Both these examples illustrate the advantages of physical proximity. "On the Internet, no one knows you're a dog," as a *New Yorker* cartoon so aptly put it. The Internet is a fine location for a chat room, but hardly a trustable source of community in day-to-day living. Proving oneself a trustworthy neighbor still counts.

Another phenomenon that will only increase in the future is *computer support for human contact.* We use technology more and more, to get hold of the people with whom we wish to connect. We joke a lot about recorded responses, "push *one* if you have a touch tone phone, push *two* if you don't." Some folks pretend to have rotary phones in the hope of getting to a human operator faster. While these call centers can be annoying some times, they cull out routine transactions so as to make help from human operators more affordable.

On the nonbusiness side, many families are beginning to put up home pages on the world wide web. Someone puts up the family tree, circulates news, organizes get-togethers, and soon, distant family members get to know more about one another. So computers can make possible an enhanced degree and quality of human contact.

Another development from the rise of technology is an *expanding range of lifestyle opportunities.* Some would argue that women remain in exactly the same economic position as that of past generations. Compare today's situation with the era of the so-called organization man. In that environment, lifetime corporate employment meant that the husband was recognized as the breadwinner. Wives couldn't sustain a career because the family had to pick up and move every three years or so as the husbands climbed the organization ladder by moving from one assignment to another.

Today, on the other hand, my daughter-in-law may sometimes start her workday in a bathrobe and bunny slippers, but she still manages to run a very successful company. Except for scheduled meetings, she can pretty much choose when to go to the office. Early mornings often suit her best, so she gets on her computer, scans overnight faxes, or returns phone calls to time zones where conventional offices are already open for work. And so she and my son, who also works for the company, are able to take care of their children in ways that would be impossible without net-worked information technology.

In my own case, the expanding range of lifestyle opportunities offered by technology allows me to telecommute from my San Francisco apartment, even though my assistant maintains her office at Lucent Technologies' New Jersey location. Together, the two of us maintain an electronically supported information bridge between Silicon Valley and Bell Labs.

Finally, the future can point to healthier *human-machine relationships*. Today, people understand computers as electronic tools, no longer as so-called electronic brains. In the past, popu-lar opinion frequently succumbed to the notion that nature's workings resembled those of the highest technology then current. Fortunately, millions of personal encounters with computer stu-pidity have undermined the grandiose claims once made in the name of "artificial intelligence."

Computers do not provide a very good role model for how to solve problems. While some people would feel complimented if somebody were to tell them they thought logically, logic is actual-ly a terrible problem-solving methodology. That's why people use it so rarely, and why computers sometimes freeze up when con-fronted with absolutely simple questions.

What Lies Ahead?

The developments that information technology has spurred are merely milestones, if you will, on a path to a future destination. I can best describe this place as a *Glass Village*, a global information *supermarket* rather than superhighway. Unprecedented transparency: everyone, and everything, is available for view via on-line packages. Much like a supermarket, because it's composed of packaged items. Cyberspace information leans heavily toward numbers, heavily toward constructs that work well with computers. Despite the rich variety, furthermore, each item bears much of its creator's imprint. For example, if someone wanted to view the weather outside via an electronic window via a remote camera, the location and orientation of each camera available for viewing has been predetermined: while there may be a large choice, each selection is prepackaged, just like in a supermarket.

In another aspect of the Glass Village, the recognition of *individual identities will be heightened to small-town levels*. Within ten years, for example, there will be a kind of "caller ID" for personal encounters. What happens when computers learn to recognize human faces?

When I go to our local dry cleaner these days, I like the fact that the person behind the counter knows who I am, because I rarely remember to bring the slips. I get my stuff and it's never a problem. For all I know the clerks might have created a crib sheet, something like: "Penzias. Balding, loquacious man. Lives up the street. Always returning from a trip." That seems okay.

But what happens if the owner decided to set up a video camera in the shop? Furthermore, suppose the system were to record the transaction, extract my name, speech characteristics, facial features, and whatever other parameters—such as height, body build, and gender—might help it to identify me on subsequent visits. What a neat memory aid for people who have trouble remembering names and faces. How about a miniature unit with an unobtrusive earphone prompter? I can see the ads now: "No more embarrassing encounters. Improve your social life! Start a new career as a *maitre d'!*" A bit scary though, isn't it?

No matter how deep and well-argued privacy concerns become, however, it is hard to imagine any combination of steps or circumstances that could block the deployment of such technology for any significant amount of time. What can be done to prevent it? All it takes is one computer whiz anxious for an improved social life and willing to post software on the Web. Like it or not, big city anonymity may well prove a momentary detour in the social history of our species. In its place, we'll all be living in small towns with potentially nosy neighbors. Hiding will take a lot of work, in a Glass Village.

Perhaps even sooner there will be an *information-glutted competition for attention.* It must be easy to create information. Just look how hard people work to give it away—on the Internet, in your mailbox, on television, even handbills when you walk down the street. Who has time to digest it all? The scarce thing is attention: all the media compete for it.

While competition among media sources seems far better than the alternatives, there is at least one unfortunate side effect. As news broadcasters, magazines, and the authors of would-be best-sellers vie with one another for audience share, they naturally look for new, different, and exciting material. But in doing so, they invariably highlight rarity at the expense of normality.

Imagine what would happen if one a student came empty-handed to class with the claim that "the dog ate my homework." A highly improbable event at best. But the perceived probability of such an event—along with a host of others—has been much enhanced by television. Because, if a dog took even the smallest nip out of a homework page anywhere in the United States, you know it would be on the next six o'clock news. In other words, the rare, the unusual, is sought out eagerly and brought to our attention. So much so, that it's very hard to see the difference between the unusual—brought to us because it attracts our attention—and the world of everyday experience that our senses encounter at first hand.

There are some good things that will happen also in the Glass Village. For example, there will be *new alternatives to congestion rationing.* In the relatively near future—probably within no more than ten years—there might be answers to the problem of cars and congestion. As networking technologies, precise navigation aids, electronic maps, and the like become available, I think we may very well see a renaissance in public transportation. Not mass transit in today's sense, but a customized public transportation system, one that can offer door-to-door service at affordable prices.

One possible service will provide the ability to call up, get a multipassenger van to arrive at your doorstep precisely when you want it, and take you to wherever you want to go, in a guaranteed amount of time. It will be so much like a chauffeured limousine that you can leave your car at home most of the time—with tremendous positive implications for general well-being. In the Glass Village, bits and bytes will stand in line for us.

And finally, the Glass Village will offer our society an *unprecedented growth in individual options and resources*. Technology acts as an enabler—and we, as human beings, are the enabled. Technology, the application of organized knowledge, by humans and for humans, will allow us to be whatever we are already—only more so.

P A R T   3

Case Studies

How have technological innovations changed the ways in which we know, we learn, and we communicate? Have they transformed the fundamental nature of these capacities? What are the likely cultural and social consequences of the major advances in communication technology?

Introduction BY ALAN TRACHTENBERG

How do the machine technologies that define modernity actually go about changing human existence? This phase of the investigation into "technology and the rest of culture" focuses on how particular technologies of communication, understood in the broadest sense as the conveyance of data from one point to another, impinge upon and alter fundamental patterns of individual and social life. In each case—printing, telegraphy and telephony, and computation (which can be taken as a form of communication within the domain of the self)—the creation of new instruments and new devices has resulted in new artifacts of cultural behavior: the reading of books, the sending of messages across vast spaces, the bringing together and sorting through of vast amounts of data in digital form.

The perspective on such changes taken by the following three case studies is that of the historian, precisely the historian as opposed to the literary critic or the political theorist or the philosopher of mind. The shared perspective here is one of inquiry into causes as much as effects, and into effects not as isolated consequences of specific machines or mechanical processes, but effects as dispersed variations in experience that can be understood as constituting the cultural meaning of these specific technologies.

What is a book? What is a telegraphic (or electronic) message? What is a telephonic conversation? What is a digital calculation? Historians seek answers to these simple-seeming but defiantly difficult questions less in the object or act itself than in the vibrating web of implication, the nuanced texture of the experience represented by these terms. Meaning lies in experience, the American Pragmatists teach, and it is to the records and accounts of experience that the historian goes. Each of these terms (book, and so on), which refer to products of technologies of mechanical (or electronic) reproduction, names a profound alteration in how people understand the nexus of relations within which they have their individual and social being: the relation between their minds and their bodies, between their minds or mind-body entities and those of others, between inside and outside, between the visible and invisible, between near and far, between space and time. Such effects of altered ways of communication within and without are incalculable. The changes occur within a continuum: from speech to writing to print to telegraphy to the computer screen. Each node represents a jolt of some degree of disturbance, perhaps a deeper disruption of the previous understanding of what it means to communicate than the collective mind realizes at the time. The historian's materials include unplanned, deflected, and unconscious effects—include, that is, the past's sense of its future as much as its beliefs about its present.

Of course historians are also interested in technology as "force," as transformative energy, as power acting within society. We want to know, and need to know, objective facts, the sort which can be measured, put into quantities of height and depth and horsepower, in units of velocity. But the less conspicuous issues that belong to the subjective realm are those that call most needfully for the historian's perspective. What does all this power, and the particular forms it takes in the all-dominating category of "communication," make of the human beings it serves? What do our machines call forth from us in the way of human possibility? Writing in 1918 about questions and conundrums much like those embedded in the topic of "technology and the rest of culture," historian Henry Adams wrote that

> the new American—the child of incalculable coal power, chemical power, electric power, and radiating energy, as well as new forces yet undetermined—must be a sort of God compared with any former creation of nature. At the rate of progress since 1800, every American who lived into the year 2000 would know how to control unlimited power. He would think in complexities

unimaginable to an earlier mind. He would deal with problems altogether beyond the range of earlier society (Adams, 1946 [1918], p. 496).

Historians serve the common good by placing such perceptions as Adams attributes to "the new American" within the horizon of history, explaining the accretion of power engendered by new technologies as the result of deliberate human decisions and describable social processes. How did new technologies get organized and distributed to people in just this form and not another? Historians understand changes in the present by reference to changes in the past, helping us see that our "past" is precisely that which explains our "present," and that what we may see only as "technology" or "the machine," is also a reflection of social power-relations, of decisions made for the benefit of some against those of others. The marketing of high technology as consumable goods seems a fulfillment of Adams's prophecy regarding the American of the year 2000. Yet that same newly empowered figure is also likely to experience an increasing loss of power when it comes to influencing the shape of public life. Adams focused on the contradiction between the power to do and the power to know. For us, a gulf seems to yawn between what we accomplish in front of our computer screens, and what we imagine we can accomplish as citizens. Has the same technology that empowers our imaginations, for example, to inhabit realms of "virtual reality," also disempowered our political will, our capacity to act collectively in the actual public realm? By viewing technology as variegated modes of collective and private experience, historians pose such questions as how it came about and what it has meant that people have learned to use books and telephones and computers, and what such accessions of communicative power might yet teach us about empowerment in the rest of culture.

References

Adams, Henry, *The Education of Henry Adams: An Autobiography* (New York: Modern Library, 1946 [1918]).

From the Printed Word to the Moving Image BY ELIZABETH L. EISENSTEIN

I was asked to discuss how the changes wrought by the advent of printing might be related to the likely consequences of current innovations in communications technology. As is so often the case, no sooner had I accepted the invitation than I began to have second thoughts. First of all, I was uneasy about the title that I'd been assigned. "From the printed word to the moving image" seems to imply that the one thing was superseded by the other—an issue to be discussed later on. Second, I am poorly equipped to deal with recent communications technology. When it comes to playing with computers, my grandchildren are more expert than am I. Third and finally, I have always tried to steer clear of speculating about the possible consequences of current developments. As is true of most historians, I am skeptical about efforts to divine the future and feel sufficiently challenged by the problematic task of understanding the past.

On the other hand, I do think historians have an obligation to place current concerns in some sort of perspective. When I saw the conference announcement that recent developments have left us "in a world dramatically different from the one inhabited by previous generations," I couldn't help thinking that this very conviction serves to link our own generation with many that have gone before. To go back no further than the 1830s, Alfred de Musset described how his generation experienced the aftermath of the French Revolution and the Napoleonic wars: "behind them, a past forever destroyed . . . before them . . . the first gleams of the future; and between these two worlds . . . a sea filled with

flotsam and jetsam . . . the present, in a word" (Musset, n.d., p. 5).
Later on, Henry Adams wrote about being abruptly cut off from
the experience of his ancestors by the Boston and Albany Railroad, the first Cunard steamer, and the stringing of telegraph
wires (Adams, 1918, p. 496). Still later, Samuel Eliot Morison said
the same thing about the internal combustion engine, nuclear fission, and Dr. Freud (Morison, 1964, p. 24).

Is the idea that a new age has dawned with the advent of new
media also embedded in our past? To place current speculations
in perspective, I've been surveying reactions to previous changes
affecting media, with a focus on developments in England and
France.

Many predictions were made after the beginning of the past
century, which saw paper-making industrialized and wooden
hand presses replaced by steam-powered iron machines. Even
while printing industries were flourishing and output was rising to
meet increasing demand, nineteenth-century observers began to
speculate that the end of the book was on hand. According to
Thomas Carlyle, the replacement of book by newspaper had
already begun in the age of the hand press, with the sharp rise in
the number of newspapers being distributed in the streets of revolutionary Paris (Carlyle, 1837, pp. 21-25). Carlyle's description of
revolutionary journalism was taken over by Louis Blanc, whose
history of the French Revolution was written after the author's
career as a journalist-turned-deputy had come to an end. In a
much-cited chapter on the emergence of journalism as a new
power in human affairs, Blanc paraphrased Carlyle. Books were
suited to quieter times, he wrote, but we are now in an era when
today devours yesterday and must be devoured by tomorrow. And
then comes the celebrated formula: the age of books is closed; the
age of the journal is at hand (Blanc, 1852, p. 122). In mid-
century also, John Stuart Mill expressed concern that most people were no longer taking their opinions from churchmen or
statesmen. Nor, he wrote, were they being guided by books. Their
thinking was being done for them by men much like themselves

through newspapers (Mill, 1947[1859], p. 66). After the century's close, Oswald Spengler summed up the gloomy prognosis: just as the age of the sermon had given way to the age of the book, he wrote, so too the age of the book had given way to that of the newspaper (Spengler, 1928, p. 463).

Taking advantage of hindsight, we may now agree that nineteenth-century observers were right to assign special significance to the emergence of a periodical press. It restructured the way readers experienced the flow of time and altered the way they learned about affairs of state (Rétat, 1985). It created a forum outside parliaments and assembly halls that allowed ordinary readers and letter-writers to participate in debates. It provided ambitious journalists, from Marat to Mussolini, with pathways to political power (Eisenstein, 1991). It gave a tremendous boost to commercial advertising. It served to knit together the inhabitants of large cities for whom the daily newspaper would become a kind of surrogate community.

Moreover, although early gazettes and newsletters had resembled books, the later dailies developed a distinctive size and format so that they had to be placed in a separate category by archivists and librarians. The front-page layout of the modern newspaper was unlike any earlier printed product. The patchwork of unrelated items containing the first paragraphs of chopped-up stories each to be continued in some other place (section B or C or D) disproved in spectacular fashion the often cited McLuhanite notion that print encouraged linear sequential modes of thought. As McLuhan himself observed, "the modern newspaper presents a mosaic of unrelated scraps in a field unified only by a dateline" (McLuhan, 1964, p. 219). Twentieth-century painters experimenting with collage techniques may well have been influenced by the front-page layout. Daily exposure to newsprint has probably accustomed successive generations to the disjunctions and discontinuities that seem to characterize much modern art and modern fiction.

But although observers were right to sense that journalism had significant transformative effects, they were wrong in assuming that the advent of the newspaper "had completely expelled the book from the mental life of the people" (Spengler, 1928, p. 461). As it turned out, book and newspaper were interdependent, their fates closely intertwined. Book sales came to hinge on newspaper advertisements and on reviews in the periodical press. Press laws usually encompassed both forms of printed output. To be sure, publishers were less likely to be prosecuted for costly volumes aimed at elites than for cheap papers that presumably stirred up the rabble. Yet efforts to control all printed output characterized authoritarian regimes in the past and still mark totalitarian regimes in the present century. Nineteenth-century liberals objected to the Index of Prohibited Books as well as to censorship of periodicals and newspapers. (A difficult book, not a readable pamphlet, has led to the recent death sentence imposed upon Salman Rushdie.)

Coexistence and interdependence were especially apparent during the age of Mill and Carlyle. For the nineteenth-century novel was often conveyed in serial form by newspapers; its chapter endings were artfully composed to keep readers in suspense until the next installment arrived. The soap opera of today and the serial novel (roman-fleuve) of yesterday had much in common. It is true that until the advent of the radio, there was nothing quite like that interruption of narratives by commercials that gave the "soaps" their name. Nevertheless, as early as the 1830s, fiction writers were complaining about the intrusion into literature of vulgar commodities for sale. In his 1834 preface to *Mademoiselle de Maupin,* Théophile Gautier expressed savage indignation at the idea of seeing his work advertised together with such items as elastic corsets, crinoline collars, patent-nipple-nursing bottles, and remedies for toothaches (p. 39).

Gautier's other complaints also strike a familiar note. The public's appetite for scandal was being so whetted by news reports of sensational trials, he wrote, that "the reader could only be caught by a hook baited with a small corpse beginning to turn blue. Men are not as unlike fishes as some people seem to think" (p. 15). As is true of television producers today, many writers expressed disgust at the vulgar sensationalism of others, but few could afford to abandon the hope of creating a sensation themselves.

Novelists were not alone in expressing concern about the effects of sensational journalism. Doctors became alarmed over the deterioration of the nation's mental health. A physician named Isaac Ray published a book entitled *Mental Hygiene* in 1863 in which he noted, among other worries, the adverse effects of crime reporting on the national psyche: "The details of a disgusting criminal trial, exposing the darkest aspects of our nature, find an audience that no court-room less than a hemisphere could hold" (Ray, p. 237).

On such issues, nineteenth-century opinions and present-day attitudes do not seem to be far apart. Although different mass media are being targeted, the complaints are much the same—which is not to say that they were or are invalid. The ubiquity of sex and violence; intrusive commercials and sycophancy to mass taste seem to present a steady-state crisis that is no less troublesome for being so persistent. The "tawdry novels which flare in the bookshelves of . . . railroad stations" offended Matthew Arnold more than a century ago (Altick, 1963, p. 310). Similar material seems no less offensive when displayed on the shelves of the airport shops of today—if, indeed there are any books placed there at all.

Still, the newspapers that are piled up in airport shops probably do not seem as threatening to book lovers at present as they did to those in the past. Disdainful remarks about sound bites often go together with respectful comments about print journalism. In view of the defects of newscasts, book and newspaper are now often coupled in nostalgic reminiscences of that golden age when print culture reigned supreme. However, librarians and archivists are less likely to be nostalgic. They still have good cause to worry

about the relentless pressure exerted by the ever-increasing output of printed materials on available shelf space.

The advent of the electronic church shows how the sermon,
once thought to be outmoded, was capable of being resuscitated.
The paperback revolution of the 1960s came as even more of a
surprise. In the present decade, chain stores opened by Barnes
and Noble and by Borders compete with Amazon, which claims to
be the world's largest bookstore and is located on the World Wide
Web. Most recently, Oprah Winfrey's television book club showed
how the use of a new medium may dramatically increase markets
for an old one. The death of the novel also seems somewhat less
likely at present than in previous years. There is even renewed
demand for nineteenth-century novels by such authors as Jane
Austen, George Eliot, and Victor Hugo, thanks to recent filmed,
televised, and staged versions of their works.

These examples may suffice to indicate that the last two centuries have witnessed not a succession of deaths—not the death of
the sermon, the book, the novel—but, rather, a sequence of pre-
mature obituaries.

In his introduction to an essay collection entitled *The Future of
the Book* (1996), Geoffrey Nunberg takes note of this phenomenon which he describes as the doctrine of supersession. This doctrine, he notes, underlies expectations (false ones it seems at the
moment) that photography would put an end to painting, movies
would kill the theater, television would kill movies. To be sure
(Nunberg does not point this out but it is worth noting), the doctrine is not always at odds with reality. The age of the hand-copied
book, like that of the horse and buggy, did come to an end. Yet,
hand-copied books were still being produced in Western Europe
more than a century after Gutenberg. At this point it should be
noted that I'm offering a Eurocentric view throughout this discussion. There are many non-Western regions that still offer
employment to scribes. Even in the West, as Curt Bühler noted
many years ago, the scribe long outlived the manuscript book and
was not superseded until the advent of the typewriter (Bühler,

1960, p. 26). One thinks of all those clerks plying quill pens in nineteenth-century law offices. And although the manual type-writer may now be on the verge of obsolescence, its keyboard, transferred to the word processor, has received another lease on life.

The advent of printing is seen to outmode not the manuscript book but the Gothic cathedral in the most celebrated case cited by Nunberg to illustrate the supersession doctrine. It comes from the chapter in Victor Hugo's *Notre Dame de Paris* where the archdeacon first points to the great cathedral and then stretches out his right hand toward a fifteenth-century printed book and announces *"Ceci tuera cela;" This* (the printed book) will kill *that* (the cathedral, which had served for centuries as an encyclopedia in stone). Nunberg does not pause over the ironic implications of Hugo's making this pronouncement while living, as he did, in the midst of a Gothic revival. Nor does he comment on the building of Gothic cathedrals in the present century—witness the Cathedral of St. John the Divine in New York and the National Cathedral, which my children watched being completed near our home in Washington, D.C. Nevertheless, he argues persuasively about the fallacy of assuming that new artifacts and styles must always supersede old ones.

Of course there are significant differences between medieval cathedral building and Gothic revival architecture just as there are between the experience of nineteenth-century readers of Hugo's original novel and that of recent viewers of Disney's *Hunchback of Notre Dame*. To complicate matters (and these issues are remarkably complex), one must also allow for the difference between the way Hugo's novel would have been received in a French-language version as against a translated one; by a nine-teenth-century reader as against a twentieth-century one. And then as bibliographers remind us, one must also allow for the way the presentation of the same text varies from one edition to another.

For printed editions do supersede each other. David Hume thought the fact that he was able continually to improve and correct his work in successive editions was the chief advantage conferred on an author by the invention of printing (Cochrane, 1964, p. 19n). Although defective early editions might be superseded by improved later ones, early editions, however defective, might also be regarded as becoming ever more valuable to rare book collectors. (Indeed, defects may even enhance the value of a printed product as in the case of a mistake in printing a stamp.) It is characteristic of our culture that markets for antiques flourish alongside demand for the latest designs. Even the horse and buggy has reemerged as a fashionable acquisition along with the antique car. Very soon, it will be the turn of the manual typewriter (but perhaps not of the mimeographing machine?).

The doctrine of supersession is much too coarse-grained to make room for such complications. Indeed, it makes no more allowance for revivals than it does for survivals. It thus encourages us to overlook what I think is most characteristic of our own era—namely the coexistence of a vast variety of diverse styles and artifacts reflecting different spirits of different times. Even the New York skyline tells the same story. Skyscrapers are certainly modern structures; yet, as others have noted, their tops bear a marked resemblance to chateaux, temples, and mausoleums. What applies to the ever-more-eclectic melange of styles and artifacts also pertains to media. That is to say, we confront an ever-more-complex mixture of diverse media: painting, woodblock, engraving, lithograph, photograph, drama, film, television, radio, video tape, walkman, phone, fax, word processor, copying machine, computer, and so on and so forth—none of which has been superseded, all of which confront us in a bewildering profusion at the present time.

The title assigned to this article, "From the Printed Word to the Moving Image," makes me uneasy because it seems to deny coexistence and implies the supersession of printed word by moving image. That the printed word is, or is about to be, superseded by something else seems most unlikely to me at present—especially when I am preparing a copy of this very article to appear in print.

Mention of preparing a copy reminds me that the photocopier has been undeservedly neglected in recent accounts. Perhaps some of you recall the television commercial for Xerox, with a monkish scribe taking a text into a monastery, reemerging with a stack of copies and proclaiming "it's a miracle" ? (This reminded me of an anecdote about Gutenberg's partner, Johann Fust, arriving in fifteenth-century Paris with a wagon load of Bibles, which the doctors of the Sorbonne then examined. Finding that each copy was exactly like every other one, they set upon Fust and accused him of black magic. The anecdote gains added resonance from the frequent misspelling of Fust's name as Faust and the resultant confusion between the legendary magician and Gutenberg's partner.) The Xerox commercial has lost ground. Newer miracles are now being hyped. Nevertheless, the copier is still indispensable to all of us who frequent archives and rare book libraries. It has dramatically changed my own working habits. I used to make sure before setting off for a library that I had pen and paper on hand to take notes and copy citations. The era of the hand-copied book had ended long ago, but the hand-copying of passages from printed books was still going strong. I recently learned that DeWitt Wallace spent hours in the New York Public Library transcribing printed passages by hand for early editions of the *Readers Digest*. Probably he developed writer's cramp as I used to do. Now, of course, I worry more about carpal tunnel syndrome. In any case, I've now abandoned pen and paper but must check to be sure I have enough coins on hand to put in the library copier. Researchers have ceased to serve as their own scribes even while they line up to endow printed pages, placed face down in a machine, with a longer lease on life.

Much as medieval universities were surrounded by stationers who farmed out pieces of texts to lay copyists for reproduction, so too late-twentieth-century universities are now surrounded by shops containing copying machines. When I was a faculty member at Michigan, it was common practice to take sections of books to the shop to be duplicated and then have the selections bound together, thus producing special anthologies of readings for certain courses. Medieval *florilegia,* common in the thirteenth century, thus reemerged in the late-twentieth century as "course packs." The publishing revolution that was set in motion by the copier has recently been arrested by lawsuits brought by publishing firms objecting to infringements on copyright and setting limits on fair use. Whatever the outcome of pending cases, the continuing struggle indicates that vital interests are still believed to be at stake in the printed word.

Of course, litigation over course packs represents only the tip of the iceberg when it comes to the destabilizing effect of very recent technologies on structures designed in an age of print to safeguard intellectual property rights. I have no idea how the control of texts by authors (and/or publishers) can be maintained in view of the floodgates that stand wide open to all the information, news, and views that are carried on the Web. Nor do I feel competent to speculate about the effects of electronic mail, the Internet, and other forms of paperless publishing on scientific research. What will happen to peer review and priority claims? I look forward to hearing from other speakers about ways of safeguarding the reward structure that has encouraged scientific innovation until now. I must confess to becoming less and less certain about the desirability of an entirely unregulated flow of information in view of those special Web sites that enable conspiratorial theorists to share their paranoid fantasies. After hearing a few snatches of hysterical commentary by people who shall remain nameless, I've even begun to question the desirability of uncensored talk radio.

Mention of talk radio brings up yet another problem about going from printed word to moving image. The spoken word is left out of consideration. I've already alluded to the revitalization of the sermon in this century. In view of the excitement generated earlier in this century about movie actors being enabled to speak, perhaps the talking image deserves attention along with the moving one. (There is also the singing star in the musical film, but I am arbitrarily setting aside all references to music and to the recording industry throughout this article). To turn back to the issue of the spoken word: although printing is silent and radio broadcasts are not, the two still have some significant features in common. In 1946, after speaking on the BBC to some twenty million people, Harold Nicolson wrote in his diary that he had no real feel for his audience. "To whom am I talking?" he asked. Although an audience of readers had been replaced by one of listeners, the sense of distance between author and invisible public remained.

Before printing, powerful lungs had been required by orators and preachers who hoped to gain a popular following. After printing, a new rather paradoxical figure emerged: the silent demagogue or the mute orator. The latter phrase was actually used to describe an influential deputy to the French Constituent Assembly in 1789. The deputy had issued an incendiary journal called the *Sentinel of the People* on the eve of the Revolution, but he was said to whisper like a woman when called on for a speech (Eisenstein, 1989, p. 193). Many of those who became prominent on the eighteenth-century political scene were notably deficient in traditional oratorical skills. In England in the 1760s, John Wilkes was an indifferent public speaker, and when he had to respond extemporaneously he fumbled his words. Tom Paine never swayed a colonial legislature with a single memorable speech. Paine's friend Brissot had a sonorous voice, but he disliked public speaking, was untrained in oratory, and timid before crowds. Camille Desmoulins stammered when he spoke. It was solely the power of their pens that gave such men a metaphorical "voice" in public affairs (Eisenstein, 1991, p. 152).

With the advent of radio and the electronic amplifier, the phenomenon media analysts call "reoralization" was greatly reinforced. Powerful lungs are still not needed (except perhaps by coaches engaged in quarrels with umpires and teachers or parents subduing noisy children), but certainly the human voice has regained lost ground. Groups gathered around radios or television sets also suggest that some of the isolating effects of individual absorption in reading materials may be mitigated. But it would be a mistake to carry this thought too far. Individual absorption in cyberspace and virtual reality is just beginning to pose new problems. A seventeenth-century writer expressed regret at the loss of conviviality in coffee houses where, he wrote, everyone now sat in "sullen silence" reading newspapers (Brewer, 1976, p. 148). One is reminded of the many fellow travellers now seen on planes or trains with earphones clamped on their heads. At least one could catch the attention of the "sullen" silent reader by making a noise, whereas nothing seems to disturb the listener wearing earphones. (Although I had planned to stay clear of the recording industry, it is too omnipresent to avoid completely: the introduction of tapes and cassettes does require more attention when considering the fate of the printed word. Perhaps another article should be entitled: "From Printed Word to Talking Book.")

But there is also the printed image to be reckoned with. After all, it is only a short step from fixed image to moving one. As a youngster I played with little books where images were arranged in sequence so that if I flipped the pages rapidly I had the illusion of watching something move. (I recently saw my six-year-old grandson playing with a similar little book he called a "flipper.") Such little books were not irrelevant to the development of animated cartoons and to the early movie industry. If we take the moving image to allude to movie and television screens then, as already noted, the newer medium not only coexists with the older one, but actually helps to boost sales of the latter. In the case of Jane Austen and company, we go from printed word to moving image and then, in reverse motion, back to increased sales of printed word.

In such disparate fields as bird watching and art history, the printed image was and still is of enormous consequence. As

recently as October 1996, a Metropolitan Museum of Art cere-
mony marked the publication of a thirty-four-volume *Dictionary of
Art,* containing fifteen thousand images and twenty-eight million
words. "Only in the age of the jet plane, the photograph, the fax
and the computer has a work like this been possible," wrote the
reviewer in *The Washington Post* (October 16, 1996, p. B8). No
mention of printed words or images. Yet this dictionary probably
owes more to cumulative results obtained by the old media than
it does to jet, fax, or computer. It represents the culmination of a
tradition that originated with Vasari's sixteenth-century illustrated
collective biography of artists—a tradition that also encompassed
Diderot's eighteenth-century *Encyclopédie,* which was subtitled *A
Dictionary of Arts and Sciences* and which contained seventeen folio-
sized volumes of text and eleven volumes of plates—eleven folio-
sized volumes, that is, devoted exclusively to pictures.

It is too often forgotten that images replicated on wood and
metal were introduced at more or less the same time as Gutenberg's invention. As William Ivins insisted, "the exactly repeatable
pictorial statement" was at least as significant an innovation as was
letterpress printing (Ivins, 1958, p. 2). On this point we ought to
follow George Sarton's advice and think of a double invention:
typography for the text and engraving for the images (Sarton,
1957, pp. 116–19). Otherwise we are likely to reinforce the mistaken notion that printing entailed a one-way movement from
image to word.

To be sure, there was such a movement, in Protestant regions at
least. As is implied in Victor Hugo's account, Bible stories presented by stone portals and stained glass went out of favor even
while Bible stories conveyed by printed chapter and verse were
being translated into vernaculars and published far and wide.
Some iconoclastic Puritans insisted on lay Bible reading while
smashing graven images.

But although newly printed Bibles and austere white-washed
churches did replace sculptured stone portals and stained glass in
some regions, in others, religious imagery was exploited by all

available means. Especially in Catholic regions, Baroque illustrations of angels, saints, and martyrs were multiplied in diverse media and circulated among the faithful as they still are being circulated even now. Nor did Puritans object to the use of printed images for didactic purposes. Indeed, picture books for children came into vogue under Protestant auspices.

Moreover, use of the printed image was by no means confined to religious, moralistic, and didactic purposes. Pornography found a large audience in sixteenth-century Europe with the publication of Aretino's verses accompanied by those graphic presentations of copulation known as "Aretino's postures." The same era saw frequent resort to political propaganda by means of printed imagery as is shown most vividly by Lutheran caricatures and cartoons. The French Revolution produced prints of peasants with pikes and the storming of the Bastille that still resonate in the modern American imagination (witness Pat Buchanan's campaign oratory). Image-driven foreign policy did not originate with television pictures of starving African children; there were newspaper wars before there were television wars. Cartoons of Belgian babies being bayonetted by brutal Germans played a part in winning support for American entry into World War I. Later, the discrediting of anti-German propaganda during World War I would encourage an unjustified skepticism about atrocities being committed in World War II.

To the historian of early modern science, probably the most important aspect of the double invention is that it led to a greater reliance on image and symbol and less reliance on words. Once it became possible to duplicate precisely rendered drawings of natural phenomena together with exactly repeatable diagrams, graphs, equations, and the like, scientific communications became less dependent on ambiguous texts whether in Greek, Arabic, Latin, or the vernaculars. Identical maps, charts, and log tables fixed on printed pages made it possible for observers located in different regions to coordinate their findings and to trace the paths taken by moving objects such as planets and comets with unprecedented precision.

The remarkable advances that were made after the discrediting of the ancient authorities, such as Galen on anatomy or Ptolemy on astronomy, help to account for the widespread acceptance of the doctrine of supersession. I'm going to sidestep current debates among historians of science about paradigm switches and simply note that to almost all nineteenth-century observers, it seemed obvious that the ancients had been surpassed in science and technology.

Among many Victorians, the doctrine of supersession (together with its counterpart, the idea of progress) was so widely accepted and fully orchestrated that it was applied to all phenomena—not just to Ptolemy and Galen or dinosaurs and dodos but to the entire course of human history and to all cultural artifacts. "In every department of life—in its business and in its pleasures, in its beliefs and in its theories, in its material developments and in its spiritual convictions—we thank God that we are not like our fathers. And while we admit their merits, making allowance for their disadvantages, we do not blind ourselves in mistaken modesty to our own immeasurable superiority" (Froude, cited by Hartwell, 1960, p. 416). I often wonder what such commentators would have made of the counter-cultural trends at work today when the march of medicine is being countered by a vogue for homeopathy and acupuncture or when reports of a moon landing are coupled with astrologers casting horoscopes in daily papers. Even now, quite a few of my contemporaries are taken aback by the resurgence of literal fundamentalism and the advocacy of "creationism" more than fifty years after the Scopes Trial in Tennessee.

Such phenomena might seem less surprising if we were not so entranced by the advent of all the new communications technologies that we failed to consider the preservative powers of print. Recently, *The Sunday Telegraph* (July 23, 1995, p. 5) announced that the Church of England was launching itself into cyberspace to enable churchmen to surf a World Wide Web of biblical information. This announcement came to mind when I saw the conference brochure refer to a trend toward globalization and assert that "the world was more homogeneous." The existence of a Web that is world-wide certainly seems to support this assertion. Its usage to spread information about the Bible, however, gives rise to other thoughts. Not only is the world still divided by adherence to different faiths, but within Latin Christendom itself Bible printing undermined the use of a single religious tongue. The Gutenberg Bible, of course, was in Latin, but the Lutheran Bible was not. Vernacular Bibles produced by Luther's followers balkanized the common Latin culture of the Western Church. New editions of modernized versions have scarcely helped to put Humpty Dumpty back together again. "English was good enough for God; it should be good enough for Texas" remarked a Texan opponent of bilingual education. Efforts to bring the Gospel to everyman are still being undertaken on a global scale and the Bible continues to be translated into hundreds of new tongues. Even now, new literary languages are being created and then fixed in print by missionary societies. The tower of Babel is growing ever higher alongside the expanding Web.

After this final example, let me offer a brief conclusion. Print culture no longer monopolizes modern communications and now shares the stage with a bewildering variety of new media. Nevertheless, the printed word has not been superseded. To understand the chaotic state of contemporary culture, we have to take into account the unsettling effects of new communications technologies. But this should not distract us from also acknowledging the continuing, ever-cumulative effects of a double invention that is now five hundred years old.

References

Adams, Henry *The Education of Henry Adams: An Autobiography* (Cambridge, Mass: Houghton Mifflin, 1918).
Altick, Richard, *The English Common Reader* (Chicago: University of Chicago Press, Phoenix edition, 1963).

Blanc, Louis, *Histoire de la Révolution Française* (Paris: Langlois et LeClercq, 1852), 12 vols., iii, Ch. 6.

Brewer, John, *Party Ideology and Popular Politics at the Accession of George III* (Cambridge: Cambridge University Press, 1976).

Bühler, Curt, *The Fifteenth Century Book: The Scribes, the Printers, the Decorators* (Philadelphia: University of Pennsylvania Press, 1960).

Carlyle, Thomas, *The French Revolution a History* (London: Chapman and Hall, 1837), 3 vols., ii, Book I, Ch. 4.

Cochrane, J. A., *Dr. Johnson's Printer* (London: Routledge & Kegan Paul, 1964).

Dudek, Louis, *Literature and the Press* (Toronto: Ryerson Press, 1960).

Eisenstein, Elizabeth L., *The Printing Press as an Agent of Change* (Cambridge: Cambridge University Press, 1979), 2 vols.

Eisenstein, Elizabeth L., "Le publiciste comme démagogue: La *Sentinelle du Peuple* de Volney," *La Révolution du Journal 1788–1794*, Pierre Rétat, ed. (Paris: CNRS, 1989), pp. 189–97.

Eisenstein, Elizabeth L., "The Tribune of the People," *Studies on Voltaire and the Eighteenth Century* (Oxford: The Voltaire Foundation, 1991), 145–59.

Gautier, Théophile, Preface (1834) to *Mademoiselle de Maupin, Texte Complet (1835)*, A. Boschot, ed. (Paris: Garnier Frères, 1966), pp. 1–39.

Hartwell, R. M., "The Rising Standard of Living in England 1800–1850," *The Economic History Review* 13 (1960–1961): 397–416.

Ivins, William, *Prints and Visual Communication* (Cambridge, Mass.: Harvard University Press, 1953).

McLuhan, Marshall, *Understanding Media* (New York: McGraw Hill, 1964).

Mill, John Stuart, *On Liberty*, A. Castell, ed. (New York: F.S. Crofts, 1947[1859]).

Morison, Samuel Eliot, *Vistas of History* (New York: Knopf, 1964).

Musset, Alfred de, *La Confession d'un Enfant du Siècle* (Paris: La Renaissance du Livre, n.d.).

Nunberg, Geoffrey, ed., *The Future of the Book* (Berkeley: University of California Press, 1996), Introduction, pp. 9–21.

Ray, Isaac, *Mental Hygiene* (New York: Hafner Publishing Co., 1968[1863]).

Rétat, Pierre, "Forme et Discours d'un Journal Révolutionnaire," *L'Instrument Périodique: La Fonction de la Presse au XVIIIe siècle*, Labrosse and Rétat, eds. (Lyon: Presses Universitaires, 1985), pp. 139–78.

Sarton, George, *Six Wings: Men of Science in the Renaissance* (Bloomington: Indiana University Press, 1957).

Spengler, Oswald, *The Decline of the West*, Charles F. Atkinson, trans. (New York: A. Knopf, 1928), 2 Vols., ii.

Shaping Communication Networks: Telegraph, Telephone, Computer

BY DAVID E. NYE

Technologies are social constructions. Machines are not like meteors that come unbidden from outside and have an "impact." Rather, human beings make many choices when inventing, marketing, and using a new device. The telephone and telegraph, the early forms of networked communication, provide an essential background for understanding the computer network. The choices made by inventors, entrepreneurs, workers, and consumers created the networked society. This society did not have a preordained form, contrary to the determinists,[1] whether they be pessimists (Ellul, 1964; Meyrowitz, 1985), or optimists (Gingrich, 1995; Negroponte, 1995). I find Langdon Winner's argument more convincing: autonomous technology is an ideology, which can and should be resisted (Winner, 1977, p. 335). A close reading of history, rather than *post hoc ergo Procter hoc* analysis, shows that human beings choose which machines they will use and how they will use them. Fernand Braudel, the great historian and author of *Capitalism and Material Life,* reflected on how slowly some societies adopted new machines, and declared, "Technology is only an instrument and man does not always know how to use it" (Braudel, 1973, p. 274). And even after people know how

125

to use a technology, they can integrate it into a society in many ways (Bijker, et al., 1987, 1992). The telegraph is a case in point.

Conceived and patented by Samuel F. B. Morse, the telegraph astounded observers in 1838. Incredulity brought excited crowds to demonstrations and the first telegraph offices often provided seating for the public, who could scarcely believe that it was possible to sever language from human presence (Czitrom, 1982, pp. 4–8). Twenty years after its invention, when thousands of miles of lines linked the states, the *New York Times* declared that, "The Telegraph undoubtedly ranks foremost among that series of mighty discoveries that have gone to subjugate matter under the domain of mind" (August 9, 1858).

But we should not write history backwards, assuming that inventors knew what was to come. They are quite commonly uncertain or wrong about how their devices will be used. Thomas Edison thought the chief use of his phonograph would be to record speech, not music (Conot, 1979, pp. 246–47). Morse did not fully understand the enormous commercial potential of his telegraph, and he tried to sell it to the federal government. While many European countries chose to make the telegraph part of the postal service, Congress refused, and it took five years before it granted Morse $30,000 to build a line from Washington to Baltimore. Yet at first "so little use was found for the original line . . . that chess games by telegraph were promoted between experts in the two cities" (Taylor, 1951, p. 152). As this incident suggests, each culture decides how to embed networked technologies within its social structure. (The government later failed to see the broadcasting potential of radio, understanding it only as a superior form of the telegraph, for point-to-point communication [Czitrom, 1982, p. 67].)

Since the government chose not to operate either the telegraph or telephone systems, private investors did. There were choices to be made. There were technical questions. Morse's telegraph relied on attentive and skilled labor to make it work. In the following decades, a small army of inventors made improvements on the original device, so that messages could be typed in on keyboards, recorded on paper, printed out, received automatically, relayed, and so forth (Israel, 1992, pp. 43–60). A good deal of inventive activity was stimulated by competition. Rival telegraph companies financed inventors such as Thomas Edison, Elisha Gray, and Alexander Graham Bell, in the search for all the possible variations that could be patented (Jenkins, ed., 1991, pp. 88–97, 109–12, 141–53).

Even as the many patents were filed, entrepreneurs had to decide whether to license their technology to others or develop it themselves. Bell lacked capital, and decided on a decentralized system, where local businessmen would build systems and make profits. The central company made its money from royalties paid by telephone manufacturers, and from leasing the telephones themselves (Smith, 1985, pp. 5–6). Entrepreneurs also had to decide whether systems should meet a minimum standard for communication or be more expensive high–fidelity systems.[2] Computer companies face the same kinds of problems today. What is the minimum level that consumers will accept, and what is the trade–off each time more bells and whistles are added? The early telephone "was a crude and limited device. To keep costs down, a fidelity level was accepted well below that at which music could be enjoyed or the phone used for entertainment. The system was optimized for its single most marketable use, conversation" (Pool, 1983, p. 28). Yet, by 1890, music was often being "broadcast" over the telephone, notably a "telephone concert" sent from New York to Rochester and Buffalo (Marvin, 1988, p. 211). Bell chose not to develop the telephone as a form of musical broadcasting, however, though the Austrians did.

There were other marketing decisions as well, some of which governments made. How accessible should a new form of communication be to the public? If Americans took a laissez-faire approach, nineteenth-century Spanish authorities retarded the spread of the telegraph and telephone, fearing they might be useful to revolutionaries. More recently, personal computers were regarded with suspicion by Soviet authorities, and in some parts of Eastern Europe even typewriters were restricted. As Ithiel de Sola Pool emphasized, "It is not computers but policy that threatens freedom. The censorship that followed the printing press was not entailed in Gutenberg's process; it was a reaction to it" (Pool, 1983, p. 226). The United States rejected press censorship, and freedom of speech was guaranteed by the First Amendment to the Constitution. But the free access to print was not maintained in the new media of the telegraph or telephone, and the monopolies that controlled them were not required to maximize the distribution of their service.

Should pricing encourage extensive or intensive use of a communications system? This is obviously a question with political implications. The Bell system during the period of its monopoly consciously decided to get a high return from a small but wealthy clientele, rather than expand the system to include the largest number of possible subscribers (Fischer, 1992, pp. 45–46). Even so, service was barely adequate, while large dividends were routinely distributed instead of being reinvested in better equipment or a larger network (Reich, 1985, p. 137). This meant that poor people and rural areas were long denied full access to this technology. For at least a generation, ordinary Americans conducted business and politics at a disadvantage. When its patent expired in 1893, Bell controlled all 266,000 of the legal telephones in the country. Its own lawyer wrote to the company president that, "The Bell Company has had a monopoly more profitable and more controlling—and more generally hated—than any ever given by any patent" (Brooks, 1976, p. 103). Nevertheless, for the next seven years the company chose to attack new telephone services in court, based on further patent claims, rather than compete on service. The independents installed no less than three million telephones in their first fourteen years, and Bell only belatedly matched this rapid expansion, offering service to many customers it had previously ignored (Fischer, 1992, p. 46). An intensive network focused on large customers was also IBM's marketing choice, until other companies challenged it with the personal computer. TelMex pursued similar restrictive, high-profit policies in Mexico until last year, and it had only five million access lines for a population of over a hundred million when its monopoly was broken up (Chapman, 1996). TelMex's level of per capita distribution then was roughly that of the United States in 1915 (Fischer, 1992, pp. 88–92).

Aside from technical and marketing decisions, each investor in a new technology has to decide when to merge and when to compete. Often entrepreneurs make what seem to be amazing mistakes in this regard. Western Union was offered the chance to buy Bell's patent rights to the telephone for $100,000, but it refused the opportunity. Instead, it developed the competing patents of Elisha Gray, hiring Edison, among others. Bell was disappointed that he could not sell his invention, and only reluctantly developed the business himself. His undercapitalized system benefited greatly from fierce competition between Jay Gould and Western Union. As a result of this conflict, Bell acquired all of Western Union's telephone patent rights in exchange for some cash and a promise to stay out of the telegraph business (Reich, 1985, pp. 132–35). Within a generation, Bell had outgrown Western Union and purchased it.

The early applications of networked technologies were commercial. The telegraph never became common in the home. This was not merely a matter of expense; most people did not have the time or inclination to master the Morse code. Telephone lines and the necessary equipment were not cheaper than the telegraph, but they were far easier to use. Nevertheless, the Bell interests viewed the phone and advertised it until the 1920s as an instrument of business and vital communication. It was not sold to the public as a means to socialize for almost two generations (Fischer, 1992, pp. 75–80). Likewise, the mainframe computer was also seen exclusively as a serious instrument of business, government, and the military for a full generation. The advent of the personal computer shifted attention to entertainment, education, and multimedia applications. Thus, there is a long trend toward democratization of access to networked communication. But even with access, people do not always know what to use the new machines for, and they seldom learn to exploit all of a device's potential.

For example, between 1838 and the middle of the next decade, businessmen did not develop the telegraph a great deal. It took time before most realized the advantages of early access to telegraphed information. When the first line connected New York, Philadelphia, and Washington, its business developed slowly. Six months after it started operation, it was taking in only about $600 a week. This meant that the line was averaging less than two thousand words a day (Taylor, 1951, p. 152). In 1845, Samuel Colt built a line from Coney Island to Battery Park, so that he could telegraph news of incoming ships to subscribing merchants. The venture was successful, and he soon operated a similar service in Boston (Robertson, 1985, p. 114). By 1848 the telegraph linked Chicago to New York, and grain traders with immediate access to information made fortunes trading in Midwestern wheat. Indeed, the telegraph made possible the national supremacy of these two cities as centers of commerce. In the 1850s, the Chicago Board of Trade and the New York Stock Exchange became national institutions because they used the telegraph to centralize buying and selling. "The wider the telegraph's net became, the more it unified previously isolated economies. The result was a new market economy that had less to do with the soils or climate of a given locality than with the prices and information flows of the economy as a whole" (Cronon, 1991, p. 121). Every small businessman in the country could send a message through Western Union, and most of the traffic on the wires was commercial.

Because of the undoubted utility of communication networks in business, they have often been touted as time-saving conveniences in the home. The sociologists studying Middletown in the 1920s thought that, "Around the telephone have grown up time-savers for the housewife as the general delivery system for everything from groceries to a spool of thread" (Lynd and Lynd, 1956, p. 173). Similarly, today the Internet is advertised as a virtual shopping mall, but is it really much more convenient than mail-order catalogs that have an 800 number? Have we really moved so far from the 1901 Sears and Roebuck Catalog? Most people still want to see and touch things before they purchase them, and Americans continue to spend large amounts of time in stores. Nevertheless, the myth that networked communication will reduce shopping has persisted for more than a century, even though Americans in fact spend increasing amounts of time at malls. The mistake here is the notion that speed is the consumer's chief concern.

Yet speed is obviously important. Before the telegraph, information traveled no faster than a horse or a sailing ship; afterwards it moved at the speed of light. Since 1838, the speed of transmission has improved relatively little, but the distances involved and the quantity and quality of what can be sent have never stopped increasing. Americans discovered that economic development hinged in many ways on the ability to communicate rapidly over thousands of miles. The telegraph line first gave corporations the possibility of keeping in constant touch with distant markets and funnelling frequent reports into a central office; it would seem that a national company could hardly function efficiently without this instantaneous communication.

However, such statements encourage us to read history backwards, assuming that people knew in advance the eventual uses of a new machine. Consider railways as an example. These were far-flung enterprises. Their employees could never be assembled in one place, and their managers had to communicate over great distances. Today it would seem clear that a technology that sped information much more quickly than the trains themselves would be of use in orchestrating their movements. Yet this was not intuitively obvious to the early railway managers, who ran "by the book," meaning that trains were expected to keep to a schedule, but could not be accounted for with any precision when they were en route. Railroads ignored the possibilities of the telegraph for fifteen years until the Erie Railroad began to institutionalize its use in the early 1850s. It did so to overcome nearly chaotic conditions, as it proved impossible to be sure where trains or their components were on the hundreds of miles of track. The telegraph enabled managers to monitor their rolling stock, and they learned to depend on it to send and receive orders, to signal when tracks were clear, or to warn trains over to a siding to avoid a collision. In retrospect it seems clear that no railroad could run efficiently without the telegraph, and yet some lines did not adopt it until after the Civil War. Even as late as the 1870s, "trains on the busy Boston & Lowell Railroad were still being run by the book instead of the telegraph at the time of a disastrous wreck" (Martin, 1992, pp. 22–24).

Once the telegraph was in use on a railway, it was possible to standardize each road's own clocks, in order to improve efficiency and service. This practice made the timetable more reliable. However, it did not immediately strike the government or the railroads as obvious that it would be a good idea to make time uniform throughout the country. A whole generation passed before the complaints of stranded passengers, caught between time systems as it were, led the railways to decide in the 1870s (more than thirty years after Morse announced his discovery) to make time uniform from coast to coast, with clearly marked time zones, and more or less the system that we know today (Trachtenberg, 1982, pp. 59–60).

Americans also used the telegraph to flash news across the continent, but this result was not foreseen in advance. Before Morse's discovery, newspapers already discovered and sold news to each other, and if a story seemed particularly important, a paper would charter a steamboat or make other special arrangements to get news out faster to the public (Blonheim, 1994, p. 48). To a considerable extent these developments were consumer driven. The Associated Press grew out of the public's intense desire, in 1846, to have news of the Mexican War. Newspapers found that it paid to make extraordinary efforts, and five New York papers soon decided that it paid even better to pool resources (Blonheim, 1994, pp. 49–51). Out of their collaboration grew the Associated Press, which soon had reporters telegraphing European news from the first ships to reach Boston (and, after 1849, from Halifax). At first other newspapers complained, but eventually most became members. While some have been tempted to propose that the telegraph promoted objectivity as an ideal standard for reporting the news, the telegraph "was superimposed on a news-gathering system that already placed a premium on apparent factual accuracy." Objectivity was a social construction, not a product of technology (Schiller, 1981, pp. 3–7).

What did nineteenth-century Americans think of this communications revolution? At mid-century the "Factories, railroads, and telegraph wires seemed the very engines of a democratic future" (Trachtenberg, 1982, p. 38). The "universal communication" of the telegraph was celebrated as a force that would help realize "manifest destiny" and bind the nation together. The *American Telegraph Magazine* declared that "nearly all our vast and widespread populations [will be] bound together, not merely by political institutions but by a Telegraph and Lightning-like affinity of intelligence and sympathy, that renders us emphatically 'ONE PEOPLE' everywhere" (Czitrom, 1982, p. 10). Likewise, *Scientific American* praised the telegraph for promoting the "kinship of humanity" (Fischer, 1992, p. 2). But the telegraph did not hinder the coming of the Civil War, and one might even argue that its rapid reports of events, such as John Brown's raid, fanned the flames of sectional conflict. Certainly it proved useful to the conduct of war, allowing generals to follow troop movements and address supply problems hundreds of miles away.

Hopes for the realization of democracy through communications technology shifted to wireless telegraphy when it was introduced in 1899 (Guthey, 1997). Power-hungry trusts could easily control lines and wires, reasoned critics, but corporations would not be able to monopolize the ubiquitous ether. Once wireless technology had evolved into home radio, commentators once again declared that communications technology would spread "mutual understanding to all sections of the country, unifying our thoughts, ideals, and purposes, making us a strong and well-knit people" (Frost, 1922, p. 18).

> Radio was portrayed as an autonomous force . . . as a technology without a history. Rarely, in those heady, breathless articles about the radio boom, was reference made to the twenty–five years of technical, economic, and cultural experimentation that had led to and produced radio broadcasting. Radio was thus presented . . . as an invention free to reshape, on its own terms, the patterns of American life (Douglas, 1987, p. xv).

The unveiling of television in the 1930s generated similar rhetoric. RCA executive David Sarnoff declared

> When television has fulfilled its ultimate destiny, man's sense of physical limitation will be swept away and his boundaries of sight and hearing will be the limits of the earth itself. With this may come a new horizon, a new philosophy, a new sense of freedom, and greatest of all, perhaps, a finer and broader understanding between all the peoples of the world.[3]

As each new communications technology appeared, the public harbored utopian expectations for it, much like the euphoria we have just passed through regarding the personal computer and the Internet.

But the repeated utopian claims for new forms of communication presaged, without exception, the creation of monopolies or oligopolies. After several decades of cutthroat competition in the telegraph industry, Western Union emerged just after the Civil War. It had not only consolidated the industry, but become one of the largest corporations in the country. Its near monopoly allowed it to set high rates. Within the telephone industry, the Bell interests predominated from the start, and they grew so wealthy that they swallowed Western Union into American Telephone and Telegraph. Manifest destiny had turned into manifest profits. There is a clear pattern here: Western Union, AT&T, the early GE-NBC radio monopoly, IBM in the 1950s and 1960s, and the dominance of the three major television networks after World War II. In each case, more competition has been forced upon these communications industries by government regulation.

While the computer network is still in rapid evolution, once again technological utopians are claiming that both a freer market and universal understanding are at hand. According to Newt Gingrich (who is echoing the Tofflers), the computer is ushering in a new era when information and expertise will be democratically dispersed to all and every citizen will be empowered and linked to every other. This so-called Third Wave will break up big business and big government, and promote greater individual autonomy. "The coming of the Third Wave brings potential for enormous improvement in the lifestyle choices of most Americans" (Gingrich, 1995, pp. 57, 7). Gingrich seems unaware of the poor track record of such claims.

One reason for the long-term dominance of AT&T, GE, IBM, and other large communications companies was their research efforts. Patents strengthened a business's position, and networked communication industries were central to the institutionalization of research. As noted earlier, Western Union and its competitors contracted with many inventors to improve the telegraph, and two of these men, Bell and Gray, later independently invented the telephone, while a third, Thomas Edison, was hired to improve it. Edison set up a permanent research laboratory at Menlo Park, using money from telegraphy invention and manufacturing. Later, AT&T founded one of the two first industrial research laboratories in the United States, the other being that of General Electric, which supplied the growing power network and was a pioneer in radio and television (Hughes, 1989, pp. 150–74). IBM also long supported a major research laboratory that kept it dominant in hardware. (It proved inept and vulnerable in software development, however, to the benefit of Microsoft.) Control of a string of new patents long assured these companies market dominance.

If networked communications did not automatically create a freer market and democratic unity, they did develop an Orwellian aspect. The telegraph or the telephone can be tapped. Computers are used to track every credit card purchase, and they are central to espionage. The wide-ranging possibilities were realized early. The telegraph made possible new monitoring and control systems: fire alarm boxes, stable calls, door bells, and burglar alarms. Each of these communicated a single message in the imperative voice: Fire! Come here! Thief! Chicago, in the decade after the terrible fire of 1871, installed 430 miles of wires to link alarm boxes and telephones to the city's fire stations (Platt, 1991, p. 18). Police installed alarms and call boxes throughout major cities. Within wealthier homes, wired communication carried the master's wishes to servants anywhere on the grounds.

Telegraph technology was also used in factories, which adopted fire alarms, watch clocks, and other control and surveillance devices. By 1875 an electric burglar alarm was available that was not merely a device attached to a vault or a door, but a system that used concealed wires and signaled the opening of doors or windows. Electric fire alarms were also available (Jerome Redding Company, 1875, pp. 4–7). Electricity appeared in monitoring systems of all sorts, most popularly in regulating night watchmen. In 1875, the "electric watch clock" was already "used in Banks, Factories, Railroad Stations, and Public Buildings, and [was] intended to show whether the watchman makes his rounds with regularity and visits the several stations at the required time." This "clock" was in fact considerably more, since it consisted of a central apparatus wired to various points on the night watchman's rounds. As he passed each point, the watchman touched a knob, which activated a marking device on the central "clock" and recorded the precise time of his signal. "The clock may be placed in the office of the superintendent. It is provided with a pencil pressing on a paper dial, upon which it makes a permanent

record of the manner in which the watchman does his duty"
(p. 8). If the watchman failed to report within a given period, the
alarm automatically sounded. By the early 1880s some alarms
were connected to fire and police departments. The secretary at
a word processor today whose every keystroke can be monitored
by managers descends directly from such night watchmen. In the
1970s, New York Telephone became infamous for its surveillance
program, in which operators and service personal were randomly
overheard by managers, to be sure that they performed according
to strict guidelines (Langer, 1970, pp. 326–33). Networked communications lent themselves to surveillance and social control.

Just as the night watchman's movements could be mechanically recorded using an electrical device, so companies began to
install time clocks, where a worker had to "punch in" and "punch
out" whenever he came and went. Once a national time system
had been invented by the railroads, being "on time" became an
absolute rather than a relative standard. Time clocks supple-
mented the use of bells and whistles to announce a plant's opening, noon hour, and closing. Here again, the machine's new
function was shaped by a preexisting desire, in this case for discipline and regularity (O'Malley, 1990, pp. 38–41). Labor historians
E. P. Thompson and Herbert Gutman have investigated worker
dissatisfaction with "factory time," which conflicted with their own
traditions of "alternating bouts of intense labor and of idleness"
over a long workday. When workers lost control over time, they
were less able to generate their own communal routine, but
instead worked according to a timetable. In the nineteenth century, workers often went on strike over this and related issues
(Thompson, 1967, p. 73; Gutman, 1977, pp. 32–43). Once again,
note that nothing in the machine itself dictates this form of social
control.

The communications industry itself became a major employer.[4]
In Chicago, for example, by the 1880s there were 175 branch
offices for the telegraph throughout the city, and the main office
alone had two thousand telegraphic instruments (Platt, 1991, p.

19). Most of these machines were operated by men, and in the first years of the telephone, men were also employed as operators. Telephone companies allegedly hired women because they were more polite to customers (Schlereth, 1991, p. 188), but just as importantly, the job did not require much technical knowledge, and women could be hired for lower wages. Over time, however, the most interesting shift is not that from men to women, but the attempt to automate as much of the business as possible. At first, all calls went through a switchboard operator, but since the rapid expansion of the 1890s, telephone companies have continually reduced the human labor needed to place calls, by using direct dialing and automatic forms of customer information and service. The Internet can extend this substitution of machines for people a good deal further, permitting automated banking, shopping, and many other services.

Yet if communications technologies can be used to automate many tasks, the human factor remains central. In the white-collar world, much routine decision-making and form-processing can be incorporated in computer software, reducing the time required and the number of employees needed. Companies that computerized often adopted a "flat structure," which eliminated layers of middle management, leading to layoffs and job insecurity among those who remain. Thus thousands of mid-level managers were fired by AT&T in the early 1990s. Yet despite many claims made for computerization, it remains unclear whether the computer really has improved white-collar productivity. A study made by the American Manufacturing Association in 1996 found that reducing staff raised profits for only 43 percent of the firms that tried it, and 24 percent actually suffered losses, despite the savings on wages. In some cases, computerization reduced the time that highly skilled employees had available to perform skilled work. "Their jobs became more diverse in a negative way, including things like printing out letters that their secretaries once did." For white-collar workers the computer often had the unintended consequence of diminishing specialization (Tenner, 1996, pp. 207–8).

As this example suggests, not all of the changes that technologies facilitate are sought by early entrepreneurs. While machines are socially constructed, this does not mean that people collectively know what they are doing. Quite the contrary, the combined choices of inventors, entrepreneurs, workers, and consumers create unanticipated cultural consequences. However, because no single individual or institution wills certain changes, they are often mistaken for attributes of the machine itself. If fundamental categories of experience change, it may seem as if technologies are responsible.

In many cases, social forces, cultural values, or business competition can readily explain a new development. For example, consider gender and the communications workforce. A wide array of studies have confirmed the common stereotype that women use the domestic telephone more often and usually speak longer than men (Fischer, 1992, p. 231). Women also have held most of the jobs at the switchboard. In contrast, the telegraph was never common in the home and was largely restricted to male operators. The main-frame computer was primarily used in science and industry, and it was for the most part programmed and controlled by men. More recently, the home personal computer and office word-processor are increasingly being used by women. These technologies are not deterministically fated for use by one or another sex. Most early telegraph manuals assumed a juvenile male audience, and a deluge of boys' books celebrated telegraph operators, inventors, and engineers. Most telegraph offices were socially constructed as male environments. Often located in the railroad station and frequented by newspaper reporters, this was no place for a lady. In addition, early operators needed mechanical skills that women were not likely to acquire in the school system of the antebellum era. These were cultural choices, not technological givens. In the case of the computer, not only did it emerge from the male-dominated fields of science and engineering, but the majority of early software games were directed at young men. Technologies are not inherently gendered. Rather, preexisting cultural values and gender relations are reexpressed in the use of new machines.

While the gendering of technologies is clearly a social process, some other sweeping claims may seem persuasive. Critics such as McCluhan, Kern, and Meyrowitz assert that new forms of communication have changed our experience of time, our perception of events, and our sense of self. While each of these subjects could easily be the subject of a study in itself, I will sketch a response to each of these three claims.

It has become a cliche of communications studies to point out that the telegraph and telephone annihilated space and time. It seems indisputable that the experience of time was fundamentally altered by instantaneity. The sense of the present, rather than being located within one's sight and hearing, suddenly expanded to include anywhere that had been wired. In *The Culture of Space and Time,* Stephen Kern devotes over one hundred pages to the discussion of time between 1880 and the end of World War I. In reaction to the establishment of a single public time and the increasing bureaucratization of the period, he argues, Victorians chose to "affirm the reality of private time against that of a single public time and to define its nature as heterogeneous, fluid, and reversible" (Kern, 1983, p. 34). Moreover, public time, even though defined as a constant, forward-moving progression, seemed to be accelerating. By 1900, Henry Adams believed that the pace of history was increasing geometrically. A myriad small acts of saving time and increasing efficiency together had created a rushing torrent of events beyond human control. Indeed, Kern argues that the July Crisis of 1914 was accelerated out of control by telegraphic reports to the newspapers, while "communication technology imparted a breakneck speed to the usually slow pace of traditional diplomacy and seemed to obviate personal diplomacy" (Kern, 1983, p. 265). Yet was this really a case of technological determinism? The Austrians and Germans forced the crisis by setting a very short deadline for Serbia to respond to their ultimatum, and they timed the presentation of their demands in such a way that it would be extremely difficult for France, Russia, and England to intervene as peace-makers. Kern's own evidence shows that the German secretary of state adroitly used the telegraph to force the crisis (Kern, 1983, pp. 262–63). Kern is correct to emphasize that the public sense of crisis was heightened by means of instant communication, but this result flows from how the technology was used, not the technology itself.

Communication is not just about information. Networks made it possible for widely separated people to share emotions. Lovers speaking over a telephone line, a father calling his children long-distance, and friends sending each other messages on the Internet are creating and sharing emotions. This is not restricted to private life. With the telegraph appeared the full-fledged media event, long before radio and television (Dayan and Katz, 1992, pp. 1–24). Even before 1838 Americans had evinced a desire to define and share experience on a large scale. When the Erie Canal was officially inaugurated in October, 1825, cannons were placed at intervals from Buffalo to Albany, and fired in sequence, starting when the first boat entered the canal. Cannon-fire communication was a slow and cumbersome process, and took several hours to convey a prearranged message. A procession of boats carrying dignitaries was feted at each town, in a journey that took weeks.[5] By contrast, huge (and spontaneous) public demonstrations greeted the news that Cyrus Field's Atlantic cable had successfully transmitted messages from Britain in 1858 (Czitrom, 1982, p. 13). Such mass participation, familiar enough in the twentieth century, had already become a regular feature of American society by the Civil War, when crowds gathered around telegraph offices for battle news.

When the first railroad across the United States was completed in 1869, a telegraph line carried the event to all parts of the country. Each blow of the hammer driving in the famous golden spike was "heard" simultaneously in New York and San Francisco. When the link was complete, celebrations began everywhere at once. Three days of largely spontaneous jubilation began in San Francisco. The crowds cheering and marching in the streets of Chicago formed one of the largest impromptu processions ever seen in that city, estimated to have been seven miles long. The news likewise brought crowds into the streets of Buffalo, St. Louis, Philadelphia, and New York (Nye, 1994, pp. 73–76). In this early example of a national experience, local festivities gained in significance because people knew that from coast to coast other people were also celebrating, and because they knew that the whole nation had been drawn together by railway and telegraphic technology. Radio and television later amplified the potential for such manifestations.[6] Networked technologies make it possible to create such media events, and they have done so since the middle of the nineteenth century. Yet media events are not inherent in the media themselves, and I would argue that Americans hunger more for them than many other peoples.

The telephone brought people closer together in the sense that distance did not inhibit a phone call, but it also had an unanticipated effect, according to some early studies: people spent less time actually in each other's presence. The Lynds found that in Middletown starting in the 1890s people began to replace visiting with telephone calls, which were briefer and less personal. A characteristic comment in the mid–1920s was: "I do very little visiting—mostly keep in touch with my friends by telephone" (Lynd and Lynd, 1956, p. 275). Today's home-based telecommuters are now voicing a similar complaint. It is tempting to use such evidence to decide that human beings are becoming progressively alienated from one another, and that machines have interposed themselves between them.

Daily experience seems to validate such an analysis, but we forget that human institutions decide to use voice synthesizers and automated answering services. Where individuals have control, they emphasize personal contacts. Personal contact remains essential in business, which explains why companies focusing on the bottom line still spend billions of dollars sending executives and salesmen out for meetings rather than rely on video-conferencing, telephone calls, or other technological substitutes. Job interviews are held in person. Most people continually telephone a small number of close friends and relatives. Even e-mail is most commonly exchanged between people who already know one another face-to-face. Rather than conclude that networked communication substitutes for personal contact, one can just as easily argue that they amplify and preserve already established relationships. This does not mean that the citizens of Middletown were wrong about the substitution of telephoning for visiting. But the telephone did not create time pressures. Rather, the telephone was used to mitigate loss of contact caused by increasing demands on people's time.

While new technologies may seem strange, over time they are not experienced as external. As people use a technology it ceases to seem foreign, and they weave it into their sense of self. In the age of steam, a metaphorical language was common that suggested people were like boilers. They needed to "let off steam" or they might "blow up." The telegraph and telephone familiarized the public with electricity, which brought with it another complex of metaphors, including such expressions as "getting your wires crossed," "a mental short-circuit," and the need "to recharge your batteries." These figures of speech were reinforced by the widespread practice of electrical medicine, in which people were literally connected to machines for mild charges of electricity, which were alleged to cure a wide range of conditions from onanism to neurasthenia. Even a "human dynamo" or a "live wire" could still "blow a fuse." These metaphorical language worlds are being overlaid today by another set of expressions based on encounters with the computer (see Sherry Terkel's paper on the computer and the sense of self elsewhere in this issue). My point is that Americans have used each new system of communications to create metaphors about their nervous systems and mental functioning (Nye, 1990, pp. 155–56). These figures of speech can also become part of a deterministic ethos, in which machines appear to control social development.

The thrust of the many examples presented here is that technology remains within human control. As Leo Marx has argued, one reason that we think it is not under our control lies in the word *technology* itself (see Marx's essay in this issue). To repeat Braudel: "Technology is only an instrument and man does not always know how to use it." Just as railroad companies did not discover the utility of the telegraph for almost two decades, we are still learning how best to use the computer. The tremendous productivity increases it was supposed to bring to white-collar work have not yet materialized, because we are still learning how to use it. The most profitable or useful ways to employ networked technologies were not obvious when they were introduced, and consumer demands helped to shape the services that each system would render. But once the nature of the service to be rendered had been defined, monopolies or dominant institutions soon formed: the New York Stock Exchange, the Chicago Board of Trade, Western Union, the Associated Press, AT&T, NBC, and later IBM. While each new form of networked communication was heralded as an agent of democratic unity and free market competition, Americans repeatedly used them to create powerful oligopolies and to increase the means of social control. Each of these institutions was eventually regulated and/or broken up by the federal government. It remains to be seen whether our use of the computer will repeat the pattern of the telegraph and telephone.

The Internet at present seems far more accessible than previous networked technologies. It allows virtually any person or group to create a home page, and permits new forms of publishing and communication. Will the Internet become the platform for essentially new forms of social relations, or will it ultimately be used to reenforce existing divisions based on class, race, and education? Will the Internet spawn new forms of national government control? Will the utopian visions of democracy and unity, which proved inaccurate for the telegraph, telephone, radio, and television, be fulfilled as people learn to use the Internet? Is Nicholas Negroponte right when he predicts that the nation state will "evaporate," replaced by an international cyberstate? Is he right to claim that "digital technology can be a natural force drawing people into greater world harmony" (Negroponte, 1995, pp. 237–38, 230). My argument here has been that such arguments are fundamentally wrong. No technology is, has been, or will be a natural force. Who is using the Internet? Governments, universities, international corporations, hackers, criminals, the CIA, and millions of individuals. The world they create with it will be no better than they are. The brave new world of the Internet will contain innovative ways to shop, learn, and communicate, but it will also contain new forms of power, fraud, misrepresentation, surveillance, and social control.

Notes

[1] I also find problematic the theory of anthropologist A. L. Kroeber, who studied the near-simultaneous invention in different places of the telephone and the light bulb and concluded that, "given a certain constellation and development of a culture, certain inventions must be made" (Kroeber, 1963, p. 150).

[2] Yet another marketing decision is whether to sell machines outright or to rent them. Bell and later IBM, when each held a near total monopoly, chose to lease their machines. This was both a profitable strategy and a means to maintain monopoly control (Smith, 1985, pp. 163–65).

[3] Horowitz, 1989, p. 178. The same kind of rhetoric surrounds digital television. The Benton Foundation has recently noted: "Digital television will be a powerful medium. Powerful enough to do some powerful things for the American people. Like using television to serve children better. Giving us political debate that really is debate. Using the new interactive and on-demand features to provide the information people want and need every day. But there's no commitment from the broadcast industry to serve these public needs." See Benton's web page on the Debate on the Future of Television, at http://www.benton.org/TV/debate.html.

4 Space does not permit discussion of the telegraph linemen, who first organized within the Knights of Labor, particularly in the western part of the United States. During the 1880s, the United Order of Linemen formed in Denver, but it never became a truly national organization. Electrical fairs increasingly brought together employees of large companies, and at a St. Louis trade fair in 1890, linemen and wiremen agreed to unionize under the aegis of the American Federation of Labor. The following year, three hundred delegates established the National Brotherhood of Electrical Workers of America (Marsh, 1928, pp. 9–24).

5 Shaw, 1966, p. 186. The *Utica Sentinel,* November 8, 1825, noted the general enthusiasm: "the whole proceeding was emphatically the work of the citizens—some classes were particularly distinguished and none felt the generous impulse more forcibly than the mechanics and military: their zeal was unbounded." The overall effect of the event was that "political animosity has for a time been banished by the generous burst of popular enthusiasm; and a whole people have poured forth the overflowing of their gratitude and mingled in a general acclamation of joy."

6 A short list of other events would include Lindberg's flight, Orson Welles's famous radio play about an invasion from Mars, Edward R. Murrow's broadcasts from London during the Blitz, the end of World War II, the launch of Sputnik, Winston Churchill's funeral, the O. J. Simpson police chase, the Gulf War, and, of course, Apollo XI (Nye, 1996, pp. 69–81).

References

Bijker, Wiebe E. and Law, John, eds., *Shaping Technology/ Building Society: Studies in Sociotechnical Change* (Cambridge: MIT Press, 1992).

Bijker, Wiebe E., Hughes, Thomas P., and Pinch, Trevor, eds., *The Social Construction of Technological Systems: New Directions in the Sociology and History of Technology* (Cambridge: MIT Press, 1987).

Blonheim, Menahem, *News over the Wires: The Telegraph and the Flow of Public Information in America, 1844–1897* (Cambridge: Harvard University Press, 1994).

Braudel, Fernand, *Capitalism and Material Life: 1400–1800* (New York: Harper Torchbooks, 1973).

Brooks, John, *Telephone: The First Hundred Years* (New York: Harper & Row, 1976).

Catalog, Jerome Redding Company, 1875.

"Celebration of the Completion of the Erie Canal," *Albany Daily Advertiser* (October 11, 1825).

Chapman, Gary, "Mexico: A Window on Technology and the Poor," *The Los Angeles Times* (October 28, 1996).

Conot, Robert, *A Streak of Luck: The Life and Legend of Thomas Alva Edison* (New York: Seaview Books, 1979).

Cronon, William, *Nature's Metropolis: Chicago and the Great West* (New York: Norton, 1991).

Czitrom, Daniel J., *Media and the American Mind: From Morse to McLuhan* (Chapel Hill: University of North Carolina Press, 1982).

Dayan, Daniel, and Elihu Katz, *Media Events: The Live Broadcasting of History* (Cambridge: Harvard University Press, 1992).

Douglas, Susan, *Inventing American Broadcasting, 1899–1922* (Baltimore: Johns Hopkins University Press, 1987).

Ellul, Jacques, *The Technological Society* (New York: Alfred A. Knopf, 1964).

Fischer, Claude S., *America Calling: A Social History of the Telephone* (Berkeley: University of California Press, 1992).

Frost, Stanley, "Radio Dreams That Can Come True," *Collier's* 69 (June 10, 1922).

Gingrich, Newt, *To Renew America* (New York: Harper Collins, 1995).

"Grand Canal Celebration," *Utica Sentinel* (November 8, 1825).

Guthey, Eric, "Newt's Clean Slate," *American Studies in Scandinavia* 29:2 (December, 1997).

Gutman, Herbert, Work, *Culture and Society in Industrializing America* (New York: Random House, 1977).

Horwitz, Robert Britt, *The Irony of Regulatory Reform: The Deregulation of Telecommunications* (New York: Oxford University Press, 1989).

Hughes, Thomas P., *American Genesis: A Century of Invention and Technological Enthusiasm* (New York: Penguin/Viking, 1989).

Israel, Paul, *From Machine Shop to Industrial Laboratory* (Baltimore: Johns Hopkins University Press, 1992).

Jenkins, Reese V., *The Papers of Thomas A. Edison*, Vol. 2 (Baltimore: Johns Hopkins University Press, 1991).

Kern, Stephen, *The Culture of Space and Time, 1880–1918* (Cambridge: Harvard University Press, 1983).

Kroeber, A. L., *Anthropology: Culture Patterns and Processes* (New York: Harbinger Press, 1963).

Langer, Elinor, "Inside the New York Telephone Company," *Women at Work: Two Classic Studies,* ed. William L. O'Neill (Chicago: Quadrangle, 1992).

Lynd, Robert S., and Lynd, Helen Merrell, *Middletown: A Study in Modern American Culture* (New York: Harcourt, Brace and World, 1956).

Mann, Donald, "Telegraphing of Election Returns," *American Telegraph Magazine* 1 (November 1852).

Marsh, Charles, *Trade Unionism in the Electric Light and Power Industry* (Urbana: University of Illinois, 1928).

Martin, Albro, *Railroads Triumphant* (New York: Oxford University Press, 1992).

Marvin, Carolyn, *When Old Technologies Were New* (New York: Oxford University Press, 1988).

Meyrowitz, Joshua, *No Sense of Place: The Impact of Electronic Media on Social Behavior* (New York: Oxford University Press, 1985).

New York Times (August 9, 1858).

Negroponte, Nicholas, *Being Digital* (New York: Vintage Books, 1995).

Nye, David E., *Electrifying America: Social Meanings of a New Technology* (Cambridge: MIT Press, 1990).

Nye, David E., *American Technological Sublime* (Cambridge: MIT Press, 1994).

Nye, David, "Don't Fly Us to the Moon: The Public and the Apollo Space Program," *Foundation* 66 (Spring, 1996): 69–81.

O'Malley Michael, *Keeping Watch: A History of American Time* (New York: Viking Press, 1990).

Platt, Harold, *The Electric City: Energy and the Growth of the Chicago Area, 1880–1930* (Chicago: University of Chicago Press, 1991).

Pool, Ithiel de Sola, *Technologies of Freedom* (Cambridge: Harvard University Press, 1983).

Reich, Leonard S., *The Making of American Industrial Research* (Cambridge: Cambridge University Press, 1985).

Robertson, James Oliver, *America's Business* (New York: Hill and Wang, 1985).

Schiller, Dan, *Objectivity and the News: The Public and the Rise of Commercial Journalism* (Philadelphia: University of Pennsylvania, 1981).

Schlereth, Thomas J., *Victorian America: Transformations in Everyday Life, 1876–1915* (New York: Harper-Perennial, 1991).

Shaw, Ronald E., *Erie Water West: A History of the Erie Canal, 1794–1854* (Lexington: University of Kentucky Press, 1966).

Smith, George David, *The Anatomy of a Business Strategy: Bell, Western Electric, and the Origins of the American Telephone Industry* (Baltimore: Johns Hopkins University Press, 1985).

Smith, Merritt Roe, and Leo Marx, *Does Technology Drive History? The Dilemma of Technological Determinism* (Cambridge: MIT Press, 1994).

Taylor, George Rogers, *The Transportation Revolution* (New York: Harper & Row, 1951).

Tenner, Edward, *Why Things Bite Back: Technology and the Revenge of Unintended Consequences* (New York: Alfred A. Knopf, 1996)

Thompson, E. P., "Time, Work Discipline and Industrial Capitalism," *Past and Present* 38 (1967).

Trachtenberg, Alan, *The Incorporation of America* (New York: Hill and Wang, 1982).

Winner, Langdon, *Autonomous Technology: Technics-out-of-Control as a Theme in Political Thought* (Cambridge: MIT Press, 1977).

Computational Technologies and Images of the Self*

BY SHERRY TURKLE

COMPUTERS offer themselves as models of mind and as "objects to think with" for thinking about the self. They do this in several ways. There is, first of all, the world of artificial intelligence research—Marvin Minsky once called it the enterprise of "trying to get computers to do things that would be considered intelligent if done by people." In the course of this effort, some artificial intelligence researchers explicitly endeavor to build machines that model the human mind. Second, there is the world of computational objects in the culture—the toys, the games, the simulation packages, the computational environments accessed through Internet connections. These objects are evocative; interacting with them provokes reflection on the nature of the self.

For many decades computers had a clear cultural identity as linear, logical, mechanistic machines. Here I tell a story of a change in the cultural identity of the computer and consequently in the kind of mirror that computers offer for thinking about the self. Computational theories of intelligence now support decentered and emergent views of mind; experience with today's computational objects encourages rethinking identity in terms of multiplicity and flexibility.

* This article draws from Sherry Turkle, *Life on the Screen: Identity in the Age of the Internet* (New York: Simon and Schuster, 1995).

I. Artificial Intelligence and Models of Mind

Artificial intelligence (AI) first declared itself a discipline in the mid-1950s. From its earliest days the field was divided into two camps, each supporting a very different idea of how machine intelligence might be achieved. One group considered intelligence to be entirely formal and logical and pinned its hopes on giving computers detailed rules they could follow. The other envisioned machines whose underlying mathematical structures would allow them to learn from experience. The proponents of "emergent" AI conceived of these underlying structures as independent agents and believed that intelligence would emerge from their interactions and negotiations. From the perspective of this second group of researchers, a rule was not something you *gave* to a computer but a pattern you *inferred* when you observed a machine's behavior.

In the mid-1960s the second, emergent model seemed as promising as the rule-driven, information-processing approach. By the end of that decade, however, emergent models had been largely swept aside in the field of professional AI. Several factors were in play. For one thing, the emergent models relied on the simultaneous interactions of multiple independent agents, but the computers of the era could only handle one computation at a time. Additionally, simple emergent systems were shown to have significant theoretical limitations, but more sophisticated mathematical techniques for hooking up parallel programs were not well developed. As information-processing models came to dominate AI, the implications for general psychology were significant, particularly since a subset of AI researchers now saw brain and computer as different examples of a single species of information-processing device.

In the late 1950s, Allen Newell and Herbert Simon, pioneers of information-processing AI, wrote a program called the General Problem Solver (GPS) that attempted to capture human reasoning and recode it as computational rules. As the GPS became well known in academic circles, some psychologists began to wonder why it should not be possible to ask similar questions about how *people* solve logical problems. In the intellectual atmosphere of the early 1960s, this train of thought was counter-cultural. American academic psychology was dominated by behaviorism, which rigidly excluded discussion of internal mental states. Orthodox behaviorists insisted that the study of mind be expressed in terms of stimulus and response. What lay between was a black box that could not be opened. So, for example, behaviorists would not refer to memory, only to the *behavior of remembering*. But computer scientists had, out of necessity, developed a vocabulary for talking about what was happening inside their machines. And AI researchers freely used mentalistic language to refer to the internal states of their computer systems—referring to a program's "thoughts," "intentions," and "goals." If the new machine minds had internal states, common sense suggested that people must have them too. Computers supported an intellectual climate in which it was permissible to talk about aspects of the mind that had been banned by behaviorism, and by the end of the 1960s, the machines had played an important role in legitimating the study of memory and inner states within psychology.

The new computationally influenced psychology for describing inner states came to be known as cognitive science. By the mid-1970s, cognitive science had been widely embraced by academic psychology, but the spread of information-processing ideas about the human mind met with significant resistance when people thought in personal terms about their *own* minds. In my own studies of popular responses to the computer in the late 1970s to mid-1980s (Turkle, 1984), I found that one common reaction was for people to agree with the premise that human minds are *some kind* of computer but then to find ways to think of people as something more. They conceded to the rule-based computer some power of reason and then turned their attention to the soul and spirit in the human machine, a position summed up in the remark of one college student: "Simulated thinking can be thinking but simulated love is never love." It was, in a certain sense, a time of a "romantic reaction" to the information-processing images of mind offered by computers. People's sense of personal identity became focused on whatever they defined as "not cognition" or "beyond information."

In the mid-1980s, a group of researchers known as connectionists presented a serious challenge to the hegemony of the information-processing approach in artificial intelligence. Connectionists claimed that the best way to build intelligent systems was to simulate the natural processes of the brain as closely as possible (Rumelhart, 1990, p. 134). They argued that a computer system modeled after the brain would not be guided by top-down rules and procedures. It would make connections from the bottom up, as the brain's neurons are thought to do. They spoke in terms of artificial neurons and neural nets. The artificial systems they described would learn by developing a large number of different network connections, and in this sense they would be unpredictable and nondeterministic. When connectionists spoke of unpredictable and nondeterministic AI, they met the romantic reaction to artificial intelligence with their own brand of "romantic machines."

The resurgence of connectionist models was closely tied to a new, more general enthusiasm about the possibilities of modeling brain processes on parallel-processing computers. By the mid-1980s, computers with significant parallel-processing capacities were becoming economically and technically feasible. Additionally, it was possible to simulate parallel-processing computers on ever more powerful serial ones. In this technical and intellectual climate, researchers in cognitive psychology, neurobiology, and connectionist AI began to think of their pursuits as more than sister disciplines but as branches of one discipline—united by the study of emergent, parallel phenomena in the sciences of mind, separated only by the domains in which they looked for them.

Information-processing AI had provided a context for experimental psychology to return to the consideration of inner process; connectionism, with its language of neural pathways designed on the template of biology, opened the way for new ideas of nature as a computer and of the computer as part of nature. It thus suggested that traditional distinctions between the natural and artificial, the real and simulated, might dissolve. Emergent systems that learned, grew, and evolved through training and experience seemed lifelike indeed.

The Appropriability of Emergent AI

By the late 1980s it was clear that even those who had been most critical of information-processing AI could be disarmed by connectionism—and in particular, its language of nondeterminism and its emphasis on learning through experience. Philosophers opined that neural networks "may show that Heidegger, later Wittgenstein and Rosenblatt [an early neural net theorist] were right in thinking that we behave intelligently in the world without having a theory of that world" (Dreyfus and Dreyfus, 1988, p. 35). In general, connectionism received good press as a more "humanistic" form of AI endeavor. Emergent theory opened possibilities for artificial intelligence to form intellectual alliances that had been closed to information processing.

Among these alliances, one of the most surprising was between AI and the psychoanalytic tradition. Freud built his original theory around the notion of drive, a centralized demand that is generated by the body and provides the energy and goals for all mental activity. Later, when Freud turned his attention to the ego's relations to the external world, he began to describe a process by which we internalize important people in our lives to form inner "objects." So, for example, the superego, conceived of as such an inner object, was formed by internalizing the ideal parent.

In Freud's work, the concept of inner objects coexisted with drive theory; we internalize objects because our instincts impel us to. But later psychoanalytic theorists were less committed to Freud's notion of drive than to the idea that the mind is built up of inner objects, each with its own history. These "object relations" theorists began to describe societies of inner mental agents—"unconscious suborganizations of the ego capable of generating meaning and experience, i.e., capable of thought, feeling, and perception" (Ogden, 1983, p. 227). For Melanie Klein, these inner agents could be seen as loving, hating, greedy, or envious; for W. R. D. Fairbairn, the inner agents could think, wish, and generate our sense of "self" from their negotiations and interactions.

During the years of the hegemony of information-processing AI, psychoanalysts had been hostile to the enterprise (if indeed, they thought about it at all). In their eyes, the premise of information processing reduced the Freudian search for meaning to a search for mechanism, as for example, when AI researchers would reinterpret Freudian slips as "information-processing errors." Emergent AI struck a different chord. Marvin Minsky's (1987) language of mind as a "society" seemed to evoke the inner agents of the object-relations school; and to some, connectionism's language of decentralized associations resonated with a French school of psychoanalytic theory that argues against the existence of a centralized, "knowing" ego.

A paper by the psychoanalyst David Olds (1994) argues a family resemblance between psychoanalysis and emergent AI. Indeed, Olds goes further and suggests that contemporary psychoanalysts *need* connectionism. For one thing, connectionism presents analysts with a way to describe the ego in terms of the brain and thus enables them to forge new links to the science of biology. For another, connectionism, which recasts the ego as an emergent and distributed system, can help analysts undermine old-fashioned views of the ego as centralized and unitary. Olds acknowledges that psychoanalysts may have a problem appropriating connectionist models because understanding the theory requires a "mathematical sophistication" that most analysts don't have. "Very few people, including most psychologists, have even a sketchy understanding" of what the theory is actually saying. But he points out that innocence of technical details has never kept psychology from mining scientific fields for their metaphors. Freud borrowed the language of hydraulics; today's analysts should borrow the language of parallel, emergent, computational intelligence.

> Many libido theorists probably did not know a great deal about steam engines; they made conceptual use of the properties which interested them. This is even more true with the early computer model; very few analogizers know a motherboard from a RAM, nor do they care. The way we *imagine* the machine handles information is what counts.
>
> The point is that what gets transferred from one realm to the other is a set of properties which we attribute to both entities (Olds, 1994, p. 590).

In the heyday of information-processing AI, Marvin Minsky justified the AI enterprise with the quip, "The mind is a meat machine." The remark was frequently cited during the late 1970s and early 1980s as an example of what was wrong with artificial intelligence. It provoked irritation, even disgust. At the time, much of what seemed unacceptable about Minsky's words had to do with the prevailing images of what kind of meat machine the mind might be. Those images were mechanistic and deterministic. Connectionism's broad appeal as well as its appeal to psychoanalysts was that it proposed an artificial meat machine made up of biologically resonant components. With a changed image of what machines could be, the idea that the mind could be one became far less problematic.

When the prevailing image of artificial intelligence was that of information processing, many who criticized the computer as a model of mind feared that it would lead people to view themselves as cold mechanism. When they looked at the computer, they had a "not me" response. Emergent AI provokes a different reaction. When people look at emergent computer models they meet the idea that the "I" might be a bundle of neuron-like agents in communication. Emergent AI offers a view of the computer that makes it easier to see the machine as kin to the human and the human as kin to the machine. The "not me" response turns into a "like me" response.

II. Experiences with Computational Objects

I have said the explicit models offered by artificial intelligence are only one of the ways in which computers influence our thinking about mind. Experiences with computational objects are another. Ten years ago, most people who worked closely with computers interacted with programs whose structure encouraged a model of thinking about the self that was linear and logical. Today, we project ourselves into a far wider variety of computational landscapes. We interact with programs, games, and simulations that present themselves as driven by evolution. And we create multiple representations of ourselves by developing personae in virtual environments on the Internet. The images of self that are evoked by such experiences are fluid and multiple, with the line between the natural and artificial more permeable than before.

In the case of the Internet, people are able to join on-line communities that exist only through computer-mediated communcation. One type of virtual community is known as a MUD (short for "multi-user dungeons" or "multi-user domains"). When you join a MUD, you create a character or several characters and specify their genders and other physical and psychological attributes. In traditional role-playing games in which one is physically present, people step in and out of character; MUDs, in contrast, offer a parallel life. Most people who spend a lot of time in virtual communities such as MUDs work with computers all day at their "regular" jobs. When they participate in virtual communities, they will periodically put their characters to sleep, thus remaining a presence in the virtual space as they pursue "real world" activities. In this way, they break up their days, "cycling through" between the physical world and a series of virtual ones. MUDs may seem exotic, but they are, in fact, illustrative of the social and psychological dynamics of most on-line sociability. The key elements of "MUDding," the creation and projection of personae into virtual spaces and the fact that you can move in and out of these spaces, also characterize the more "banal" forms of on-line community such as "chat" rooms.

"Cycling through" in MUDs and other virtual environments is made possible by the existence of what have come to be called "windows" in modern computing. Windows facilitate a way of working with a computer that makes it possible for the machine to place you in several contexts at the same time. As a user, you are attentive to only one of the windows on your screen at any given moment, but in a certain sense, you are a presence in all of them. When writing a paper in bacteriology, for example, you are "present" to a word processing program on which you are taking notes and collecting thoughts, "present" to communications software that is in touch with a distant computer for collecting reference materials, and "present" to a simulation program that is charting the growth of bacterial colonies when a new organism enters their ecology. Each of these activities takes place in a window and your identity on the computer is the sum of your distributed presence. In practice, windows have become a potent metaphor for thinking about the self as a multiple, distributed system. According the this metaphor, the self is no longer simply playing different roles in different settings, something that people experience when, for example, a woman wakes up as a lover, makes breakfast as a mother, and drives to work as a lawyer. The life practice of windows is of a distributed self that exists in many worlds and plays many roles at the same time.

This notion of the self as distributed and constituted by a process of "cycling through" undermines traditional notions of identity. Identity, after all, from the Latin *idem*, literally refers to the sameness between two qualities. On the Internet, however, one can become many, and one usually does. If, traditionally, identity implied oneness, life on today's computer screen implies multiplicity and heterogeneity.

A Case Study: Self States and Avatars

Case, a thirty-four-year-old male graphics designer, plays a series of female characters in MUDs whom he describes as "Katharine Hepburn types—strong, dynamic, 'out there' women." He says that they remind him of his mother. "She says exactly what's on her mind and is a take-no-prisoners sort." Case describes his father in different terms, as a "Jimmy Stewart type." His parent's style has left its legacy: he sees assertive men as bullies but assertive women please him. Case says he likes MUDding as a female because it makes it easier for him to experiment with assertiveness both on-line and off.

> There are aspects of my personality—the more assertive, administrative, bureaucratic ones—that I am able to work on in the MUDs. I've never been good at bureaucratic things, but I'm much better from practicing on MUDs and playing a woman in charge. I am able to do things—in the real, that is—that I couldn't have before because I have played Katharine Hepburn characters.

For Case, life on the screen provides what the psychoanalyst Erik Erikson called a "psycho-social moratorium," a central element in how Erikson thought about identity development in adolescence (1963). Although the term *moratorium* implies a "time out," what Erikson had in mind was not withdrawal. On the contrary, the adolescent moratorium is a time of intense interaction with people and ideas. It is a time of passionate friendships and experimentation. The moratorium is not on significant experiences but on their consequences. It is a time during which one's actions are not "counted" in quite the same way as they will be later. In this context, experimentation becomes the norm rather than a brave departure, facilitating the development of a "core self," a personal sense of what gives life meaning that Erikson called "identity."

Erikson wrote extensively about the life cycle in terms of stages of development, but never in the spirit of suggesting stages as rigid sequences. He was well aware that with incompletely resolved stages people simply move on, trying to do the best they can. They use whatever materials they have at hand to get as much as they can of what they have missed. Case's ability to use on-line life to work through issues about assertiveness and gender identity illustrates how cyberspace has come to play a significant role in the life cycle dramas of self-reparation. Time in cyberspace reworks the idea of the moratorium because it may now exist on an always-available computer window.

Case tells me his Katharine Hepburn personae are "externalizations of parts of myself." In one interview I use the expression "aspects of the self," and he picks it up eagerly, for MUDding reminds him of how Hindu gods could have different aspects or subpersonalities, all the while having a whole self. Case's gender-swapping has enabled his inner world of aspects to achieve self-expression, but in his view, without compromising the values he associates with his "whole person." In response to my question, "Do you feel that you call upon your personae in real life?" Case responds:

> Yes, an aspect sort of clears its throat and says, "I can do this. You are being so amazingly conflicted over this and I know exactly what to do. Why don't you just let me do it?" MUDs give me balance. In real life, I tend to be extremely diplomatic, nonconfrontational. I don't like to ram my ideas down anyone's throat. On the MUD, I can be, "Take it or leave it." All of my Hepburn characters are that way. That's probably why I play them. Because they are smart-mouthed, they will not sugarcoat their words.

In some ways, Case's description of his inner world of actors who address him and are capable of taking over negotiations is reminiscent of the language of people with multiple personality disorder (MPD). But the contrast is significant: Case's inner actors are not split off from each other or his sense of "himself." They hold no secret knowledge; they do not need to be isolated. On the contrary, Case experiences himself very much as a collective self, not feeling that he must goad this or that aspect of himself into conformity. To use Marvin Minsky's (1987) phrase, Case feels at ease in his "society of mind."

Objects-to-Think-With

Appropriable theories, ideas that capture the imagination of the culture at large, tend to be those with which people can become actively involved. They tend to be theories that can be "played" with. So one way to think about the social appropriability of a given theory is to ask whether it is accompanied by its own objects-to-think-with that can help it move out beyond intellectual circles.

For instance, the popular appropriation of Freudian ideas had little to do with scientific demonstrations of their validity. Freudian ideas passed into the popular culture because they offered robust objects-to-think-with. The objects were not physical but almost-tangible ideas, such as dreams and slips of the tongue. People were able to play with such Freudian "objects." They became used to looking for them and manipulating them, both seriously and not so seriously. And as they did so, the idea that slips and dreams betray an unconscious started to feel natural.

In Freud's work, dreams and slips of the tongue carried the theory. Today, life on the computer screen carries theory. In on-line life, selves are made and transformed by language and the notion of a decentered identity with multiple aspects is concretized by virtual experiences. In making this claim I am not implying that on-line life is causally implicated in the dramatic increase of people who exhibit symptoms of multiple personality disorder, or that people on MUDs have MPD, or that MUDding (or on-line chatting using a series of different user personae) is like having MPD. What I am saying is that the many manifestations of multiplicity in our culture, including the adoption of multiple on-line personae, are contributing to a general reconsideration of traditional, unitary notions of identity. On-line experiences with "parallel lives" are part of the cultural context that supports new theorizations about multiple selves.

In thinking about the self, *multiplicity* is a term that carries with it several centuries of negative associations. Contemporary social theorists tend rather to describe a *flexible* self, an idea that can serve as an intellectual Trojan Horse for smuggling in an acceptable notion of multiplicity into the discussion of normality (Gergen, 1991; Martin, 1994; Lifton, 1993). Their flexible self is not unitary, nor are its parts stable entities. A person cycles through its aspects and these are themselves ever-changing and in constant communication with each other. Daniel Dennett (1991) speaks of the flexible self in his "multiple drafts" theory of consciousness. As when several versions of a document are open on a computer screen, a user is able to move among them at will. Knowledge of these drafts encourages a respect for the many different versions and imposes a certain distance from them. Donna Haraway (1991a), too, argues the salutary side of a split self. For Haraway, a "split and contradictory" self is a "knowing self." She is optimistic about its possibilities: "The knowing self is partial in all its guises, never finished, whole, simply there and original; it is always constructed and stitched together imperfectly; and *therefore* able to join with another, to see together without claiming to be another." Both Dennett and Haraway are describing split selves where (in contrast to the fragmentation of multiple personality) the lines of communication among the aspects of the self are open. The open communication encourages an attitude of respect for the many within us and the many within others.

In the psychoanalytic tradition, too, there is an effort to use the trope of "flexibility" as a way to introduce nonpathological multiplicity. Analysts are trying to think about healthy selves whose resilience comes from having access to their many aspects. Philip Bromberg (1994), an analyst who writes in this tradition, argues that our ways of thinking about "good parenting" must shift from an emphasis on confirming a child in a "core self" toward helping a child develop the capacity to negotiate fluid transitions between self-states. For Bromberg, the healthy individual knows how to be many but also knows how to smooth out the moments of transition. He says: "Health is the ability to stand in the spaces between realities without losing any of them—the capacity to feel like one self while being many" (Bromberg, 1993, p. 166).

Cyborg Doubles and Cy-dough-plasm

On-line life is not the only manifestation of the computer culture that encourages ideas about identity in terms of multiplicity and "cycling through." Computational objects such as toys and simulation games are also playing this role. One place to see their impact is in the ways children use them to construct new notions of what it is to be "alive."

In Jean Piaget's classic studies of the 1920s on how children thought about what was alive, the central variable was motion. Simply put, children took up the question of an object's "life status" by asking themselves if the object could move of its own accord. When in the late 1970s and early 1980s I studied children's reactions to a first generation of computer objects that were physically "stationary" but that nonetheless accomplished impressive feats of cognition (talking, spelling, doing math, and playing tic-tac-toe), I found that the focus had shifted to an object's psychological properties when children considered the question of its "aliveness" (Turkle, 1984). The presence of computational objects disrupted the classical Piagetian discourse about aliveness, but the story children told about computational objects in the early 1980s had its own coherency. Faced with intelligent toys, children took a new world of objects and imposed a new world order, based not on physics but on psychology.

By the mid-1990s, that order had been strained to the breaking point. Children will now talk about computers as "just machines," but describe them as sentient and intentional. When they talk about the life status of computational objects, children cycle through a discourse about evolution as well as psychology and physics. They resurface old ideas about the relationship between life and physical motion in terms of the communication of bits across a network. Faced with ever more complex computational objects, today's children act as theoretical tinkerers. The make do with whatever materials are at hand, whatever theory can fit a prevailing circumstance.

My current collection of comments about "what is alive" by children who have played with small, mobile robots and with computer programs that evoke evolution (such as the games of the "Sim" series and a computer simulation known as Tierra) includes the following notions: the robots are in control but not alive, would be alive if they had bodies, are alive because they have bodies, would be alive if they had feelings, are alive the way insects are alive but not the way people are alive; the digital creatures in Tierra are not alive because they are just in the computer, could be alive if they got out of the computer and got onto America Online, are alive until you turn off the computer and then they're dead, are not alive because nothing in the computer is real; the digital creatures in the "Sim" games are not alive but almost-alive, they would be alive if they spoke, they would be alive if they traveled, they're alive but not "real," they're not alive because they don't have bodies, they are alive because they can have babies, and finally, for an eleven year old who is relatively new to SimLife, they're not alive because these babies don't have parents. She says: "They show the creatures and the game tells you that they have mothers and fathers but I don't believe it. It's just numbers, it's not really a mother and a father."

In the short history of how the computer has changed the way we think, it has often been children who have led the way. For example, in the early 1980s, children—prompted by computer toys that spoke, did math, and played tic-tac-toe—disassociated ideas about consciousness from ideas about life, something that historically had not been the case. These children were able to contemplate sentient computers that were not alive, a position that grownups are only now beginning to find comfortable. Today's children are taking things even further; they are pointing the way towards a radical heterogeneity of theory in the presence of computational artifacts that evoke "life." Different children comfortably hold different theories, and individual children are able to hold different theories at the same time.

Additionally, today's children speak easily about factors that encourage them to see the "stuff" of computers as the same "stuff" of which life is made. A nine-year-old showed an alchemist's sensibility in describing "shape shifting": "In the universe, anything can turn to anything else when you have the right formula. So you can be a person one minute and a machine the next minute." From the youngest ages, children play with "transformer toys" which can be configured as machines, robots, or animals (and sometimes as people). Children play with these plastic and metal objects and in the process learn about the fluid boundaries between mechanism and flesh.

I observed a group of seven-year-olds playing with a set of transformer toys that can take the shape of armored tanks, robots, or people. The transformers can also be put into intermediate states so that a "robot" arm can protrude from a human body or a human leg from a mechanical tank. Two of the children were playing with the toys in these intermediate states (that is, in their intermediate states somewhere between being people, machines, and robots). A third child insisted that this was not right. The toys, he said, should not be placed in hybrid states. "You should play them as all tank or all people." He was getting upset because the other two children were making a point of ignoring him. An eight-year-old girl comforted the upset child. "It's okay to play them when they are in between. It's all the same stuff," she said, "just yucky computer 'cy-dough-plasm.'" This comment is the expression of a cyborg consciousness as it expresses itself among today's children: a tendency to see computer systems as "sort of" alive, to fluidly "cycle through" various explanatory concepts, and to willingly transgress the boundaries between the natural and the artificial.

In his history of artificial life, Steven Levy (1992, pp. 6–7) suggested that one way to think about artificial life in relation to the traditional concept is to envisage a continuum in which a computer simulation that demonstrates evolutionary properties would be more alive than a car but less alive than a bacterium. My observations of how children are dealing with their new computational objects-to-think-with suggests that they are not constructing hierarchies but are heading toward parallel definitions, which they alternate through rapid cycling. Parallel definitions, like thinking about one's identity in terms of parallel lives, gets to be a habit of mind.

Adults, too, use the strategies of parallel definitions and cycling through to think about significant aspects of self, although they do so with far more discomfort. In *Listening to Prozac* (1993), Peter Kramer, a psychiatrist, writes of an incident in which he experienced this discomfort. After prescribing an antidepressant medication for a college student, the patient appeared at the next therapy session with symptoms of anxiety. Kramer was not concerned since it is common for patients to respond with jitters to the early stages of treatment with antidepressants. Sometimes the jitters disappear by themselves, sometimes the prescriber changes the antidepressant or adds a second, sedating medication at bedtime. As Kramer explained this to his patient, the patient corrected him. The patient "had not taken the antidepressant. He was anxious because he feared my response when I learned he had 'disobeyed' me" (p. xii).

> As my patient spoke, I was struck by the sudden change in my experience of his anxiety. One moment, the anxiety was a collection of meaningless physical symptoms, of interest only because they had to be suppressed, by other biological means, in order for the treatment to continue. At the next, the anxiety was rich in overtones . . . emotion a psychoanalyst might call Oedipal, anxiety over retribution by the exigent father. The two anxieties were utterly different: the one a simple outpouring of brain chemicals, calling for a scientific response, however diplomatically communicated; the other worthy of empathic exploration of the most delicate sort (p. xii).

Cycling through different and sometimes opposing theories has become how we think about our minds, just as cycling through different aspects of self has become a way of life as people move through different personae as they jump from window to window on their computer screens.

Today's adults grew up in a psychological culture that equated the idea of a unitary self with psychological health and in a scientific culture that taught that, when a discipline achieves maturity, it has a unifying theory. So when adults find themselves cycling through varying perspectives on themselves ("I am my chemicals" to "I am my history" to "I am my genes"), they, like Kramer, often become uneasy. But today's children are learning a different lesson from their computational objects and on-line experiences. Donna Haraway characterizes irony as being "about contradictions that do not resolve into larger wholes . . . about the tension of holding incompatible things together because both or all are necessary and true" (1991b, p. 148). In this sense, today's children, growing up into irony, are becoming adept at holding incompatible things together. They are cycling through cyberspace and cy-dough-plasm into fluid and emergent conceptions of self and life.

References

Bromberg, Philip, "Shadow and Substance: A Relational Perspective on Clinical Process," *Psychoanalytic Psychology* 10(1993): 147–68.

Bromberg, Philip, "Speak That I May See You: Some Reflections on Dissociation, Reality, and Psychoanalytic Listening," *Psychoanalytic Dialogues* 4:4 (1994): 517–47.

Dennett, Daniel, *Consciousness Explained* (Boston: Little Brown and Company, 1991).

Dreyfus, Herbert, and Dreyfus, Stuart, "Making a Mind Versus Modeling the Brain," in Stephen Graubard, *The Artificial Intelligence Debates: False Starts, Real Foundations* (Cambridge: MIT Press, 1988).

Erikson, Erik, *Childhood and Society*, 2nd rev. ed. (New York: Norton, 1963 [1950]).

Gergen, Kenneth, *The Saturated Self: Dilemmas of Identity in Contemporary Life* (New York: Basic Books, 1991).

Haraway, Donna, "The Actors Are Cyborg, Nature Is Coyote, and the Geography Is Elsewhere: 'Postscript to Cyborgs at Large,'" in Constance Penley and Andrew Ross, eds., *Technoculture* (Minneapolis: University of Minnesota Press, 1991a).

Haraway, Donna, "A Cyborg Manifesto: Science, Technology, and Socialist-Feminism in the Late Twentieth Century," in Donna Haraway, ed., *Simians, Cyborgs, and Women: The Reinvention of Nature* (New York: Routledge, 1991b).

Kramer, Peter, *Listening to Prozac: A Psychiatrist Explores Antidepressant Drugs and the Remaking of the Self* (New York: Viking, 1993).

Levy, Steven, *Artificial Life : The Quest for the New Frontier* (New York: Pantheon, 1992).

Lifton, Robert Jay, *The Protean Self: Human Resilience in an Age of Fragmentation* (New York: Basic Books, 1993).

Martin, Emily, *Flexible Bodies: Tracking Immunity in American Culture from the Days of Polio to the Age of AIDS* (Boston: Beacon Press, 1994).

Minsky, Marvin, *The Society of Mind* (New York: Simon and Schuster, 1987).

Ogden, Thomas H., "The Concept of Internal Object Relations," *The International Journal of Psycho-Analysis* 64 (1983): 227–41.

Olds, David, "Connectionism and Psychoanalysis," *Journal of the American Psychoanalytic Association* 42 (1994): 581–612.

Piaget, Jean, *The Child's Conception of the World*, Joan and Andrew Tomlinson, trans. (Totowa, N.J.: Littlefield, Adams, 1960).

Rumelhart, D., "The Architecture of Mind: A Connectionist Approach," in Michael I. Posner, ed., *Foundations of Cognitive Science* (Cambridge: MIT Press, 1990).

Turkle, Sherry, *The Second Self: Computers and the Human Spirit* (New York: Simon and Schuster, 1984).

Turkle, Sherry, *Life on the Screen: Identity in the Age of the Internet* (New York: Simon and Schuster, 1995).

PART 4

Science

How does technology transform science and how do the sciences transform the goals of technology?

Introduction BY NICHOLAS HUMPHREY

$\mathbf{M}_Y$ grandfather used to vex me by praising my brother for "thinking with his hands." Even as a small boy I knew that the default assumption was that a person thinks with his head and that the rest of his body could be considered relatively dumb. When Rodin's *Thinker*, for example, rested his bronze head on his bronze hand, it was obvious in which of these two bronze masses the bronze thinking was meant to be going on. Yet here was my grandfather complimenting my brother for doing his thinking the wrong way round.

The three papers in this section address the relationship between technology and a special kind of thinking: science. If theory can be considered head and technology hands, then these papers have in common their emphasis on how much of scientific thinking too is done with the hands. Indeed they make it clear how far—to a surprising degree—in the history of science the theorizing head has tended to be the pupil rather than the master of the working hands.

It has not always been so clear. The picture many of us inherited from an earlier phase in the philosophy of science was more one in which theory was always in the lead. Pure science took precedence over applied science. Theorists talked down to mere technicians. The best science was motivated by intellectual curiosity, and if practical applications potentially followed from it their development could and should be left to the backroom professionals. The theory of relativity, for example, had been for Einstein a work of purely abstract speculation, even if later an army of technicians would turn his insight into an atomic bomb. The discovery of the genetic code had been for Watson and Crick a thrilling exercise in reading a chemical cipher designed by evolution, even if later a whole industry of genetic engineering would be raised on it.

Yet it hardly needs saying now that this picture bore almost no resemblance to reality. For a start, as several papers in this volume show, the fact is that technological development has mostly occurred quite independently of science. Throughout most of human history, the invention of new tools and new techniques has owed little if anything to theory of any kind at all. Instead, in the fields we now call manufacture, agriculture, medicine, warfare, and so on, people have usually solved the practical problems of how to get things done more by good luck than good theoretical management—relying on trial and error, guesswork, copying (and fortunate mistakes in copying), and the selective retention of improvements—without their understanding or even asking why the eventual solutions worked. Indeed, the modern pattern of theory-led technological invention, which now seems to many of us typical, began to emerge only around two hundred years ago.

But that picture of theory being ahead of practice was unrealistic in another way as well. For not only has the development of technology until recently depended hardly at all on scientific theory, in fact scientific theory has depended very heavily on innovations in technology. It may be true that when Faraday in 1840 was asked by Queen Victoria if she might see his laboratory at the Royal Institution, he asked his assistant to go and fetch it—on a tea tray. But, if so, Faraday was just showing off. The reality is that he and most of his fellow scientists would have got nowhere at all without the increasingly sophisticated tools that technology provided for them. Again and again, new theoretical insights have in fact had to wait on the invention of new instruments of observation, measurement, and manipulation: better clocks, better lenses, better centrifuges, and so on.

Nor does the role of technology in science stop there. For theoretical insights have had to wait also on the invention of new tools of thought: and, as Peter Galison describes in the riveting paper that follows, these tools of thought may sometimes have been modeled on tools of practice. Everyone knows how Watson and Crick relied for their theoretical breakthrough on the newly available X-ray diffraction photographs of DNA; but not everyone knows how Einstein may have relied for some of his key ideas about relativity on images that were put into his head by his experience as a patents' clerk reviewing applications for new gadgets and practical inventions.

Still, it would be a mistake to swing too far the other way, and to replace the old picture of theory dominating practice with an exaggeratedly post-modern picture of practice dominating theory. In reality, the relationship is complex and reciprocal—with the emphasis shifting one way, then the other.

I am reminded of the renewed debate in academic psychology about how best to describe the relationship between body and mind. For most of psychology's history, the accepted picture has been the rationalist one, originating with Plato and urged further by Descartes, of there being two separate entities, body and mind. All higher intellectual faculties were assumed to belong to the queenly mind, and the body was relegated to the status of a lumpish hand-maiden. But there is now a revolution under way. As psychologists get closer to appreciating how minds and bodies really interact, they are being forced to recognize how far the body itself participates in the very processes of decision-making and intelligence that were formerly considered exclusively mental. Indeed, they are coming to see not only that the idea of a disembodied mind no longer makes theoretical sense, but that the body itself must be considered an integral part of the mind: being wise, intelligent, and even knowledgeable in its own right.

We all, it seems, think with our hands besides doing it with our heads and both together. As the three papers in this section illustrate, we do science the same way. Maybe the deep reason for this common pattern is precisely that human thinking has a fractal structure, with the same pattern of interaction between head and hands emerging at every level—cultural as well as individual—at which we engage the world.

Technology and Culture

BY MARVIN MINSKY

W E'RE always trying to understand why people do the things they do. It would help us to know much more about the machinery inside our brains. But lacking that knowledge, we're forced, instead, to invent simpler explanations. This often leads to ideas like these, to describe and explain our psychology:

- A basic "survival instinct" makes us try to avoid fatal accidents.
- A basic "pleasure principle" forces us to seek enjoyment.
- Our aversion to discomfort drives us to avoid annoyances.
- Our emotional systems try to reduce our conflicts and agitations.
- Our superegos obey (or oppose) our introjected parents' goals.

Each such idea has virtues and deficiencies. For example, that survival instinct hypothesis seems to "explain" much of what we do—but its basic idea is wonderfully false, because its cart is pushing its horse when it suggests that there's one central force that empowers all those behaviors. During evolution, our brains accumulated hundreds of different mechanisms, each of which serves in particular ways to protect us from certain kinds of harm. That "survival instinct" idea makes it harder for us to see why all those different "protective" effects could come from natural selection. To understand how those systems work, we'll have to examine them one by one—instead of assuming some vital force that has no explanation.

177

Here's another "pop theory" of mind: we try to do the things we like, and tend to avoid the ones we abhor. Of course we like to do things that we like, so this statement at first seems vacuous. But "liking" is really much more complex, as every athlete and thinker discovers when they learn to find pleasure in discomfort and stress. In this essay, we'll pursue this dumbbell idea. I'll posit it here in a challenging form in order to get your attention.

This view of "culture versus technology" is not the popular one in which "culture surpasses technology because of its deeper, more human, and more spiritual character." That's the view in which artists are more esteemed than nerds, boldness of imagination is venerated over critical understanding, and feeling is felt to be more genuine than thinking. (My own view is that it's a category error to regard emotion as different from thinking. Instead, each emotion is merely a different way to think. The conventional view—of emotion and thought as opposites—is a myth that I think we may have evolved to keep us from understanding ourselves.) In any case, we're now moving into a new phase of life that will offer two great new alternatives.

Culture and Technology

When a culture can't change what people don't like, its thinkers find ways to change how they think. A good way to do this is making up myths. Suppose someone wondered what caused "it" to rain. Well, perhaps it was some supreme being's idea—so let's call that being the Rain God. Now we can make more sense of things! If it's the case that you don't like rain, perhaps the Rain God is punishing you—and maybe there's something that you can do. Perhaps you didn't pray enough or, possibly, you prayed the wrong way. Perhaps you can think of some sacrifice, or even try to mend your ways. Religions embody prepackaged replies to questions that had no good answers when their doctrines and creeds were compiled.

At any point in history, there always remain things we can't yet control. But we still have the option of dealing with them, not by changing the world, but by changing our minds. Each culture invents its arts, myths, and religions to explain what we don't yet understand—or to "justify" what happens to us. We use cultural myths to justify the values and goals that we've evolved, and our methods for making decisions. Unless we embrace those simplified rules, we'd always be facing problems that are incomprehensibly complex; they help us think that things make sense. Religions, especially, try to explain the "meanings" of unpleasant accidents. They do this by using a wonderful trick that helps to maintain and strengthen their myths. That trick—called "faith"—is supported by an insidious form of self-reference: if you make yourself hold those beliefs, then you're sure to earn huge rewards—but otherwise, you'll seal your fate. It's a really grand version of carrot and stick, for suppressing your cravings for evidence.

The other approach to things that you don't like is to try to prevent them from annoying you. Invent an umbrella. Understand meteorology. Develop effective technology. Learn more about how things actually work. It helps to be able to plan ahead—to predict what might happen for each available action. But it rarely suffices to know, just by rote, which actions worked in the past, because no two situations are ever the same. That's where the methods of science come in—that is, making up theories and trying them out.

Thus, culture and science-technology, albeit in their different ways, both help us try to make sense of things. Science keeps testing and growing and learning; it welcomes new theories whenever they're shown to work better than the previous ones. Some cultures tend to get "set in their ways" and become more conservative. When science arrives on the scene too late, when a culture already has frozen ideas, then leaders may choose to fiercely oppose new theories and experiments.

We're always meeting people who say, "Your worship of science is just one more faith." They don't understand why science is not just another collection of old cherished myths. Truly, it stands at an opposite pole; it's the very antithesis of faith. Science strives to overthrow old fables, legends, tales, and creeds—by trials, tests, and experiments. It's true that many laymen think that the scientists are conservative, with vested interests in keeping faith with their old ideas. "Why else would those scientist so blindly reject our stories about strange incidents?" If scientists treasure anything, it's overturning old beliefs! But other people don't understand that astounding claims need more evidence.

Why do so many cultural myths persist in spite of reliable evidence? There are good psychological reasons for this. We love the myths that make us feel good, and we know that if we test those beliefs we're in danger of discovering things that could weaken our faith and make us feel worse. Nevertheless, we've learned a lot over the past few thousand years. We learned to use tools, language, fire, agriculture, and domestic animals. Then came writing, metals, printing, wireless, and computers. These helped us control not only the world, but also ourselves and each other. In cases where we're able to, we try to modify things we don't like by exploiting new technologies. When we learned enough about animals and vegetables, then we could hunt—and domesticate—reliable supplies of food, labor, and materials. When we learned enough about materials, geometry, and mechanics, then we could make clothing, shelters and machines.

I've mentioned inventions that extended our powers to control things in the physical world. But now, we're seeing new kinds of inventions that could alter our lives in other ways. Computers are bringing us to the verge of creating entirely different, new worlds—worlds in which things are (completely) controlled. Those are the worlds that we call today "virtual reality."

Virtual Reality

Everyone's seen computer games. You turn your home computer on—and find yourself in an alien world. In the early years of computer games, that world might include a few object-like things—but we'd also meet, from time to time, some simple autonomous beings with whom we'd make some very simple relationships. We'd react to those creatures, and they to us, but all those encounters were violent and brief: your goal was to get from A to Z, but you'd keep finding monsters in your way—and then, if you could, you would kill them. These duels were always to the death. Eventually you'd miss your mark, and suffer a sudden and horrible fate. (But in contrast to reality, you still could continue to play the game—by paying a fee for another new life.)

Soon the art of designing those "games" evolved in more complex directions. The goal became finding your way through the world, to find and acquire some Holy Grail—and to do this you'd have to deduce from sparse clues where you should go to find it. The monsters became more dreadful, too, but they also became more clever. Now the way for you to win was no longer just being the quickest to shoot, but by solving some difficult problems. Such adventures could take hours to solve, during which you had to gain knowledge and skills. Most adults regarded those kinds of games as useless, escapist fantasy—while many children found those worlds to be loaded with significance. Just as in the real world, unexpected things keep happening—but, unlike the real world, each new event has some meaning which, if you grasp it, soon might save your life.

These games are still in their early days. Many researchers are working now to endow their creatures with more resources. At first those monsters knew only enough to play the roles of threats and traps. Now the designers are making them do more of the things that human minds do; some of them have already evolved into useful and helpful servants. Soon they'll begin to play the roles of sympathetic comrades and friends, and eventually we'll see them acting as teachers, mentors—and leaders. As their programs increase in knowledge and ability, these virtual worlds will grow more engaging than reality.

Most people regard the things in those worlds as inferior versions of physical things. But this assumes that there's some special virtue in reality. Is this really the best of all possible worlds? Must we accept it, just as it is, and not try to think of improvements? Instead, one could take an opposite view: yes, the objects in those virtual worlds are not very much like real things; they're highly oversimplified. But that doesn't make them inferior. Consider that they never wear out. They never malfunction, break, or die— unless they've been purposely programmed to. We can think of those objects not as imitations, but as "ideals"—that is, as perfect, abstract realizations of what we'd want real things to do. Consider that a real chair has no idea that it's a chair. It has no idea about what it's for. A chair can wear out, or come loose at the joints. When you use it too hard, it is likely to break. This cannot happen in virtual worlds, because virtual chairs simply don't know how. We're used to regarding simulations as approximations of things in the world. But why not think, instead, of real things as imperfect versions of what they should be. The artifacts in those virtual worlds are flawless versions of abstract ideas.

You can buy a computer program today that can simulate children's building blocks. You can use it to build towers, arches, and viaducts—the same things you could do with real blocks. But you can also use that program to construct an enormous cathedral! As a child, I tried to build such things with blocks, but always something would go wrong. You'd bump the structure with your foot, and then have to start all over. Eventually, when you ran out of blocks, you'd have to compromise your goal, and adopt a less ambitious one. Finally, when you're all done, you have to waste time at cleaning things up. In the virtual world, you won't run out of blocks, because your computer has room for millions of blocks. In the virtual world, a single command will store your structure in a memory so you can continue your work another day; you'll never have to start over. As for all that cleaning up, all you need to do is type "quit"—and all your mess will go away.

The virtual realities of the present day still are not quite real enough. You can see and hear what's going on, but few systems yet let you touch and feel. We need more research in the domain of tactile and haptic feedback. Very few people work on this now but eventually, when its value is seen, this will be a major industry.

Art and Illusion

Artists have always produced illusions. That's their job. But the artifacts that they create don't really do what they seem to do. A portrait of a suffering dog does not have any pain in it, no matter that its viewer may feel more anguish than she ever felt before. The characters in a fiction-book never actually move or think or feel—no matter what those lines of text cause to happen in their readers' minds. Listen to Marcel Proust: "Each reader reads only what is already inside himself. A book is only a sort of optical instrument which the writer offers to let the reader discover in himself what he would not have found without the aid of the book." Here Proust has used an analogy to induce you to think in a certain way. His words exploit the rhetorical trick of not specifying which "instrument." That way, he makes your mind decide if it's a telescope or a microscope. (Indeed, it could be merely a mirror, although, when you come to think of it, as Richard Feynman did in his book *QED,* a mirror is really not simple at all, because it must perform an astonishingly intricate transformation that operates, simultaneously, on all the members of an at-least-five-dimensional manifold of wave-elements.)

Thus, writing (and all the other arts) employs highly developed technologies to make their mindless artifacts evoke meanings in their audience. Still, those effects are illusions. Previous artists had no technologies for creating real characters. Now we're approaching the threshold of ways to make artifacts that can think for themselves. Soon—at least on historical time-scales—we'll be able to make things that think and feel in ways resembling what humans do. Who will design the minds of those things? What will we call those professions? Will that be called art or technology? Who cares? Nothing but trouble is likely to come from loading old words with new meanings.

Designing those new "virtual minds" will require new technical tools. We can scarcely imagine how they'll work. To construct a human-like character will need the engagement of vast libraries of resources, organized by using the techniques of artificial intelligence (AI) research. How will we represent and "modularize" what once were called "mental faculties"? Composing a new person's character may become the greatest new art of the future.

We'll also want to use those techniques for ourselves, as well as for virtual worlds. Wouldn't you like to acquire new talents, without the pain of learning to use them? Why should it take so many years to learn to play the violin, or to understand mathematics, or whatever you'd like? Once we learn how learning works, we'll surely be able to speed it up, through new connections to our brains. What will be the result of all this, when talents become commodities, and anyone can edit copies of their personalities?

The Future

Everyone's heard the common joke that nothing's certain but taxes and death. But when nanotechnology comes of age, we'll have all the energy we need, and all the required materials. And when artificial intelligence matures, no one will need to work at all, because all the material things we need can be provided by robots. Furthermore, we could use those same techniques to repair all the damage that age does to our cells—so that then we'll all live in perfect health, for as long as anyone wants to.

What then will remain for people to do? How could anyone spend so much time? I've asked many people if they would like ten thousand more years of healthy life. Most nonscientists seemed dismayed at this thought, fearing that it would end in monotony. But I found no such doubts in my scientist friends, all concerned with problems so hard and deep that no foreseeable span of time might be long enough to solve them. Could this show another difference between our culture and our technology?

Three Laboratories*

BY PETER GALISON

1. Hierarchy and Homogeneity

IN all too many realms of the history of technology and science, historians have reenacted the social relations of the disciplines about which they write. For many years, scientists looked down on engineers as "merely applying" their more fundamental, basic, central, high-level, and abstract work. This view was canonized in philosophical, historical, and scientific literature that raised the higher social station of pure science to an epistemic priority. Science, it was said, stood at the top of a Comtian pyramid of knowledge, with technology at the base; ideas trickled down from the "pure" to the "applied," technology "spun off" from the enterprise of unfettered inquiry. Spun off, trickled down, or applied: whatever the metaphor, the message was the same. The order of the world put science before technology. Over the last twenty years, work in the history of science and technology has begun to redraw this picture, blurring the boundaries between realms, and insisting on a more reciprocal (and more interesting) relation.[1]

What interests me most is one half of this move against hierarchy: that is, I am intrigued precisely by a picture in which scientific and engineering cultures are treated on a par, but not in any way homogenized. More specifically still, I have wanted to use the history of the laboratory to probe the changing nature of the interaction between the technical-industrial and scientific

* This study builds on the study of Albert Einstein's patent work developed in Galison, *How Experiments End* (Chicago: Chicago University Press, 1987), chapter 2, and on the analysis of the bubble chamber and Time Projection Chamber analyzed in greater depth in Galison, *Image and Logic: A Material Culture of Microphysics* (Chicago: Chicago University Press, 1997), especially chapters 5, 7. I would like to thank Deborah Coen for assistance in the preparation of this essay.

cultures.[2] Of course one could proceed otherwise, examining, for example, the myriad cases in which bits of scientific knowledge do enter into the production of economically significant objects: tracking quantum physics into the fabrication of transistors, lasers, or nuclear weapons. What is striking is not the manifest circumstance that physicists aided in the design of a beam or a bomb, but that the design of machines significantly altered the conception and development of cyclotrons, particle detectors, and even certain paths to the arcane reaches of renormalization in quantum field theory (see Leslie et al., 1985; Galison and Hevly, eds., 1992; Kevles, 1995; Needell, 1983; Schweber, 1988; Forman, 1987).

2. Laboratories and Material Culture

I find it helpful to begin thinking through the relation of physics and technology by focusing on specific sites—laboratories—in which instruments and experimental practice come face to face with technological structures. In a necessarily schematic way, I'd like to carve up the twentieth-century laboratory into three epochs, and in each to draw us in from the wider culture to a specific address to explore the changing relations between science and technology. Such an approach, of necessity, is not one in which one takes various theories (say nonrelativistic quantum mechanics) and seeks to trace the tortuous path by which this knowledge intersects with particular technologies (say lasers or transistors). Instead, by attending to laboratory practices themselves we can see scientists and engineers as they sought, at different times in the last century, to find a culturally specific form that their joint work might take.

For our present, very rough purposes, let's divide up the long twentieth century into three periods. The first, "modernist" period ranges from the mid-nineteenth century to World War II, a period characterized by the development not only of thermodynamics and the steam-powered technologies that both boosted it and followed from it, but also by the electrical industry with the technologies and theories that accompanied it. The second period is shorter—the period of hot and cold wars ranging from the extraordinary growth of radar-electronics and nuclear fission in the early 1940s through the 1970s. This was the period in which big physics made its initial mark: the origin of the factory-laboratory surrounding the massive accelerators, electronics, and associated technologies that emerged from the technical materiel of World War II and the Cold War. For reasons I will outline in a moment, it might be helpful to think of this industrialized physics as late modern. Finally, from the 1970s through the present—and perhaps for some time ahead—we might think about physics as organized differently, that is, neither around the "modernist" axes of steam and electricity, nor around the "late modernist" technics of "nucleonics" (as it was referred to at the time) and electronics. Instead, what characterized much of the last twenty-five years of "fundamental physics" (that problematic term) is an increasing reliance on highly sophisticated data processing at every stage: from design, to simulation, to experimental runs, to the processing of data. Hybrid devices, spatially dispersed analysis of data, and massive collaborations make particle physics of the late-twentieth century resemble not so much the factory of the 1960s as the multinational corporation of the 1990s. As difficult, contentious, and irritating as a term like *postmodern* can be, there are aspects of postmodernism, especially architectural postmodernism, that are quite useful in understanding the novelty of what has happened and is happening in high-energy experimental physics.

3. High Modernism: The Mechanical and Electrical Laboratory

Nineteenth-century physicists working on the constitution of matter spoke about the mechanical world view, a vision—a scientific promissory note, as it were—that the phenomena of ordinary mechanical objects, but also of light, heat, electricity, and magnetism, would all find their explanation in mechanical principles. Some thought that electric and magnetic fields would turn out to be stresses and strains in an infinite "world ether" that was every-where. Others took "mechanical explanation" to mean just that the mathematical structure of Newtonian mechanics would work equally well to formulate the laws of, say, electricity and magnet-ism. Yet others hoped and expected that mechanical machines could be constructed that would be analogous to—that is obey the same laws as—heat, light, and electromagnetism. Granting that the term *mechanical* carried a somewhat blurred reference throughout the nineteenth century, it was still utterly plausible for a nineteenth-century physicist to have held that the world was, through and through, mechanical. Imagine looking around: every steam-powered factory, steam-powered ship or train embod-ied and reasserted the triumph of mechanics. Gears, entrain-ments, governors: the world of salient objects, the powerful new means of transport and production coming into existence throughout Europe and the United States, stood out against the older and familiar parts of the world. Elaborate, precision-built, and powerful mechanisms represented a public theater of mechanical triumph, and for physicists like Maxwell and Boltz-mann, their theories showed that these mechanical principles could be extended in fine down to the atoms that made up gases, out to the orbiting particles that formed Saturn's rings, and deep into the very ether that carried the vibrations of electric waves (see Jungnickel and McCormmach, 1986; Harman, 1982; Buch-wald, 1985).

Not surprisingly, much of physics of the late-nineteenth century circulated around such mechanical questions: working out the science of thermodynamics, including the second law, which said no heat engine is ever perfectly efficient and entropy always increases. Problems of a practical sort refined theoretical notions of energy, entropy, and heat; these ideas fed back into a techno-logical domain. These ideas, crafted and refined through appli-cation and theory, also propelled scientific-philosophical debates that worked themselves out on both sides of the academic/public divide: the irreversibility of natural processes and the "heat death" of the universe were two.

But I promised the specificity of a laboratory. Consider Charles T. R. Wilson's cloud chamber, perhaps the greatest single scientific instrument of the first half of the twentieth century. Known to physicists for almost the entirety of this century, the cloud chamber, with its distinct tracks tracing the paths of particles, made the abstractions of atomic and nuclear physics appear as visually evident as skywriting. Along with related images, these became the canonical symbol of physics itself. Wilson himself was a quintessential Victorian scientist. Born in Scotland, he spent his youth not only fascinated by the highlands, with its extremes of weather and atmospheric phenomena, but also with their reproduction. While visiting the weather station at the harsh summit of Ben Nevis, Wilson surely encountered the devices of another meteophile, John Aitken, who had designed a dust chamber to capture the tiny dust motes that he took to be the nucleating origin of rain and snow on the one hand, but also vectors of the catastrophic diseases afflicting industrial London, and modern cities more generally.[3] Wilson modified Aitken's dust device to show how clouds of condensation could form without dust—purely, he argued, using the invisible ions that floated through the atmosphere. But like Aitken, Wilson wanted to make natural phenomena *in vitro;* Wilson meant it literally when he called the device a "cloud chamber." He would master the beautiful and eerie effects of halos and Brocken spectra and other more arcane atmospheric occurrences. Soon, however, he turned his meteorological miniatures into something else, an image-making machine that would reveal the "cloud tracks" of particles. Alpha particles, beta particles, atomic and nuclear scatterings became events visualizable with a device that could be fashioned by a schoolchild; in the hands of the best of the experimentalists, it became an instrument that helped produce the field of particle physics, revealing the muon, the heavier cousin of the electron, the "strange particles," and striking confirmation in 1939 of the new process of nuclear fission.

Does science shape technology or technology science? In an important sense, the question itself misleads. What we have in an instrument like the cloud chamber is an intersection point, a zone of exchange between the ear-shattering thunderstorm and the most rarefied theory of matter.

But if the cloud chamber was a quintessential Victorian mechanical instrument, it was born when the economics of mechanical (steam) power were already ceding to electricity. Berlin became what one historian has called the "Elektropolis," and residents of almost any major European city would have seen their environment changing almost daily in the years after 1900. In Germany, Allgemeine Electrizitätsgeselleschaft (AEG) was powering huge industrial applications—everything from the extraction of aluminum to streetcar systems, not to speak of the 248 generating stations they had already put on line (Hughes, 1983, p. 178). Berliners saw the streets carved up to support the electrical Stadtbahn; companies like Siemens and Halske or the AEG not only powered the extraordinary growth in electrical industry, but also some of the institutions that bound industry to science. Siemens himself was instrumental in getting the state to fund the Physikalisch-Technische Reichsanstalt (PTR) (Cahan, 1989). Within the walls of a quintessentially "modern" institute like the PTR, a problem could participate in the world of industry and that of science simultaneously. Black-body radiation—the distribution of the different frequencies of light by a heated object—is a case in point. To physicists like Max Planck, Wilhelm Wien, and Albert Einstein, the theory behind these glowing wires led straight to the center of radiation, matter, and eventually to the quantum discontinuity. To electrical industrialists, black-body radiation was altogether practical—how would the filaments of lights glow, with what frequencies, what intensities, at what temperatures?

At the Palais de l'Electricité, which opened in the portentous year of 1900 in Paris, the public could see this new world of electricity celebrated and displayed in apparatus, from generators to telephones (Bennett et al., 1994). Physicists joined in the electrophilia: the American physicist Henry Rowland launched a company to promote his new multiplex telegraphy, the German physicist Walther Nernst began investigations into a better electrical light.

Architecture not only reflected the new technologies, it grew out of them. Famously, the AEG building designed by Peter Behrens came to stand for modernist streamlining. Spokesman for the most scientific (even scientistic) of the modernist city-transformers was Hannes Meyer, successor to Gropius at the Dessau Bauhaus. In his famous manifesto, "The New World" (1926), Meyer captured the technomodernism that was so strikingly grounded in the products of the mechanical and electrical innovations:

> The "Norge" has circled the North Pole! Completed! The Zeiss Planetarium at Jena. To be Tested Today: Flettner's rotoship! These were, in 1926, the most recent exploits in the mechanization of our globe. Striking examples of exact calculation, they verify the ever-widening scientific penetration of the world around us. Wherever one looks, the diagram of the present shows the straight lines and mathematical curves of mechanical and scientific knowledge. . . . Airplanes stream through the atmosphere, Fokker and Farman increase our mobility and our elevation above the earth. . . . The spark of electric signs, the blare of loud-speakers, the tooting of auto-horns, the shrieking colors of advertisements . . . their simultaneous occurrence widen our conception of "Time" and "Space" without measure."

"Already," Meyer wrote in modernist zeal, "our children scorn the puffing steam locomotive and, as a matter of course, coolly entrust themselves to the electrical marvels of the Diesel engine."

Along with the children, physicists too scorned the mechanical in favor of the electrical. In a famous article of 1900, Wilhelm

Wien had kept the older, mechanical quest for unification, but inverted world views: instead of explaining electromagnetism in terms of mechanics, he would have physics explain ordinary mechanical laws—such as inertia—in terms of the behavior of the electromagnetic properties of the elementary constituent of electricity.

This was the world that young Einstein entered, engaged and transformed; a physics universe perched precariously between two imperial, unifying, and modern world views of the mechanical and the electrical. He saw the parade of electrical progress at close quarters, every day. Not long out of Zurich's Eidgenossische Technische Hochschule, Einstein took a job as patent examiner at the Swiss patent office from 1902 to 1909. Day after day, patent applications would cross his desk. Not only would he have a ringside seat for the next generation of technology, his job was, over and over, to isolate the essentials, identify the relevant physical and design principles, and strip away all that was superfluous. In 1907, for example, he blasted a proposal for an alternating current machine jointly submitted by Berlin's powerful AEG and Bern's Naegeli and Company. It was, he asserted, "incorrectly, imprecisely, and unclearly prepared" (Einstein, 1995, letter 67). Nor was Einstein content to evaluate others' patents. In a steady stream of letters he corresponded with his friend Paul Habicht both about Paul's patent ideas and about his own. Their correspondence leaped between flying machines (1908), telephone amplifiers, and Einstein's own idea for an electrostatic generator that he called his *Machinchen* (see Einstein, 1995, docs. 67, 86, 93, 124, 134, 161). For decades, Einstein's other inventions brought him a minor, if constant income (Hughes, 1993; Pyenson, 1985; Flückiger, 1974).

But Einstein's most important original technical work was launched when he was called in 1915 to testify as an expert witness in a patent litigation between an American company and a German company over the originality of a gyrocompass. Crucial for circumpolar ship navigation (magnetic compasses failed near the poles), for large ships with electrical systems that scrambled magnetic compasses, and for airplane guidance systems where the accelerative forces rendered magnetic compasses notoriously unreliable, the gyrocompass promised to become an essential piece of navigational gear. Not only did Einstein isolate the central physical principles of the case, he invented his own variation on a compass that was, in fact, later produced by Anschutz-Kaempfe. Even more importantly, thinking about the compass led him to a novel account of ferromagnetism, one that he then checked in the laboratories of the Physikalisch-Technische Reichsanstalt (see Galison, 1987, Ch. 2).

Einstein's modernism was, if you will, a modernism of stripping down, with a constant emphasis on extracting the essential and deleting all that was decorative, superfluous, or in the physics, metaphysical. Would this invention violate the first law of thermodynamics (conservation of energy) or the second law (no *perpetuum mobile*)? In the case of the contested gyrocompass patent, Anschutz-Kaempfe was suing Elmer Sperry's company for infringement. Sperry responded by displaying an earlier compass that, he claimed, did the job just as well as Anschutz-Kaempfe's— a standard patent defense. Einstein's verdict was as devastating as it was concise: the number of degrees of freedom through which the earlier compass could rotate was too small, so the compass would precess wildly under the oscillation of the ship. Compass useless, case closed. Thinking about science viewed through a young patent officer's critical eye, we can imagine Einstein differently than he is usually portrayed. We can view his electromechanical work in the patent office not as a mere sidetrack to his "real" career (a career to be understood as the blossoming of Einstein as philosopher-scientist), but rather as a constitutive phase of the modernist Einstein as examiner-physicist. Kant and Hume surely, but also Siemens and Anschutz.

That is: we can come to see Einstein learning his pragmatism, his search for basic principles, and above all his constant return to the simple principles of devices as issuing from his front row seat to the play of modern technology. Perhaps, only half tongue in cheek, we should describe Einstein as a modern patent engineer of physics, his theory of space and time not just a philosophical brand of neo-Machian operationalism, but a patent officer's simple, but brilliant intervention. Space and time: here is their machinic representation. Lay out a rigid ruler and displace identical clocks to be coordinated by light signals. Drop the ether as an idle gear; the device of nature doesn't need it. Einstein the patent-examiner interrogated the objects around him by modern, minimalist standards, criteria not different from those he so rigorously applied to the elegant kinematical foundations of physics itself.

4. Late Modernism: The Nuclear and Electronic Laboratory

I have suggested that the laboratory of high modernism turned around the twin axes of steam and electricity. And while it is hard to imagine Albert Einstein as "typical" of anything at all (who else, after all, can be said to have toppled three hundred years of Newtonian gravity?), his engagement, professionally and spiritually, with the technology of his day was startlingly representative—from electrical generators, heat engines, navigation devices, unipolar dynamos, to electrically driven refrigerators. But practically nothing in the Einstein-like consulting pattern, even the small businesses of prewar physicists, prepares us for the weapons-based physics factories of World War II and the Cold War decades that followed.

In the hills above Berkeley, during 1940 and 1941, the cyclotrons of Ernest Lawrence's laboratory—prewar big physics—forced two additions to the periodic table: Neptunium and Plutonium. Highly fissionable though it was, plutonium was hardly ready for immediate inclusion in nuclear weapons: the world's plutonium supply produced in Berkeley was so infinitesimal that it wouldn't register on a scale until the summer of 1942 (Heilbron et al., 1981, p. 42). Made by irradiating uranium by squadrons of neutrons, and letting the resulting isotopes go through beta decay, the new element could be chemically isolated, a far easier process than the physical isotope separation project that had been assigned to Oak Ridge, Tennessee.

To provide the bath of bombardment of neutrons, one had to have a nuclear reactor. So reactors, having first been demonstrated at Chicago under Enrico Fermi's direction only on December 2, 1942 (Segré, 1970, pp. 128ff.), now, in much expanded form, would become the plutonium cauldron, replicated and integrated into the largest war construction project that the United States undertook. Indeed, in the two years after Fermi's pile went critical, some 137,000 construction workers passed through the Hanford, Washington site; this Manhattan Project town was built to accommodate over 6,000 workers and their families. Each of the two separation sites was built on two thousand acres—the "cell buildings" alone were eighty-five feet high and 800 feet long and 100 feet wide; the laboratory as a whole cost some $350 million, about 16 percent of the $2.2 billion Manhattan Project (Sanger, 1989, pp. 47–55). A similar scaling-up of physics was taking place from coast to coast in a myriad of projects. From the radar laboratories of MIT to Johns Hopkins work on the proximity fuse, from the rocket project at the California Institute of Technology to the uranium separation factory at Oak Ridge, physicists found themselves in the middle of scenes like those of Hanford. No longer outside industry looking in, physicists began, in the midst of the war itself, to imagine a postwar world in which the physics laboratory would look more like Hanford than the PTR.

As for industry, David Hounshell has shown in some detail how closely Du Pont, which ran Hanford, modeled their production of plutonium on their earlier manufacture of nylon. Both were originally made as tiny laboratory samples; both had to become the output of massive factories (Hounshell and Smith, 1988; Hounshell, 1992). Part of Du Pont's "industrial culture" was a view about how this upscaling should take place, a view that took as sacrosanct the necessity of organizing research into projects, and supervising those projects by a central steering committee. Led by the engineer and project manager, Crawford H. Greenewalt, who had also run the nylon triumph, intermediate stages for plutonium production were established as the company sought to expand their operations. These so-called semi-works mediated between the test tube and the production line. For many physicists on the Manhattan Project, such stepwise increments appeared time-consuming luxuries that waylaid an expansion program that couldn't help but work as one moved up in size. If physical law conflicted with engineering practice, Eugene Wigner and his colleagues insisted, it was the engineers who had to yield. Declassified portions of Greenewalt's wartime diaries capture the difficulty he had in coping with the Chicago physicists. From his perspective, the first difficulty was his *lack* of power over the scientists with whom he had to work.

> The arrangement I'm to work under is most difficult. I have no authority over the Chicago crowd—but am to see to it by diplomacy and pleading that they do the right things at the right time and don't chase too many butterflies. What an assignment! Fortunately Compton is a swell guy so it may work out O.K. (Greenwalt, Diaries, December 19, 1942).

More contact over the next few days only reinforced Greenewalt's assessment of the quasiautonomy of the two communities, an autonomy that manifestly disturbed a man used to a more univocal authority structure of a large industrial concern. As he put it on December 22, 1942, "It is obvious that I couldn't successfully 'boss' the physicists. This can only be done by Compton for whom they all have a great respect. I outlined for him some of the major points on which we need data promptly" (*Diaries*).

Nor was this friction purely generated by the combination of two independent hierarchies. The physicists clearly had their own sense of the character of their efforts as manifestly distinct from that of the engineer. Greenewalt wrote, "C[ompton] has peculiar ideas as to the difference between 'scientific' and 'industrial' research. I started some missionary work to convince him that the difference was more of terminology than actuality but will have to do more" (*Diaries,* December 28, 1942).

Not surprisingly, Greenewalt's missionary work never succeeded. Despite his protestation that there was merely a verbal difference between the groups, a few days later he was reeling with anger at the inadequate leadership and structure of the physicists' work on the air blast cooling mechanism for the pile. "No mechanism yet devised for unloading and sorting, no flow sheet, operating manual or program. No clear idea as to what Du Pont is expected to do,—Hell! The first thing to do is to work out an operating organization. . . . I believe we *must* infilter pile design in spite of the fact that we aren't very welcome" (*Diaries,* December 28, 1942).

From the other side, Eugene Wigner, Enrico Fermi, and Leo Szilard all wanted to infiltrate the Du Pont design structure, insisting that the engineers should bring even the details of design plans to them, because (as Greenewalt put it) "some small point might violate physical principles" (*Diaries,* January 5, 1943). While the physicists continued their exploration of the nuclear physics of the pile, especially the calculation of the multiplication constant k, Greenewalt was worrying about finding group leaders with sufficient industrial research experience, "freezing" the pile design before procurement and development plans were set, and initiating work on fabrication techniques. Again and again, Greenewalt recorded the clash of cultures: "more arguments with Wigner and Szilard on transfer of information" (*Diaries,* January 6, 1943). Wigner got angry enough to bring his complaints to President Truman; Greenewalt, for his part, insisted on the difference between laboratory conditions where "everything works well" and the results at the semi-works. Indeed, Greenewalt insisted that physicists calculate under *operating* conditions, underlining the word *operating*—as in his mandate that physicist Alvin Weinberg take into account the real operating temperature of the reflector when calculating the temperature dependence of that crucial factor k (*Diaries,* July 29, 1943; August 12, 1943).

Under the pressure of total war, physicists and industrialists had to work together; many, like Wigner, were deeply persuaded that a delay might cost the Allies the war—should atomic explosives fall into Hitler's hands first. Yet struggles over identity were not matters of mere vanity. At stake was the very meaning of the terms *physics, physicist,* or *experiment* as these new factory-scale machines came on line in Hanford, Oak Ridge, Chicago, and Los Alamos. A discipline useful to industry had itself become industry and at a scale comparable to the efforts of a General Motors or Boeing, with comparable grounding in funding and power. By war's end, physicists approached "concepts of matter" in a very different way. Their instruments were different, their methods of calculation had shifted, but most of all the physicists themselves were nearly unrecognizably transformed.

Take Enrico Fermi: before the war, he had run table-top sized experiments testing the effects of ramming neutrons through paraffin into thimble-sized targets of various elements. During the conflict, he was in charge of building the first nuclear reactor; within a few years of Hiroshima, he was sitting on the General Advisory Committee, the highest council to the Atomic Energy Commission, and ultimately to the president on all matters nuclear. J. Robert Oppenheimer, Hans Bethe, Edward Teller, John Wheeler, and many others made similar transitions. No longer inventive individuals consulting industry or government on matters for which they held specific expertise, physicists began to take on industrial-scale enterprises themselves.

Luis Alvarez learned the war-industrial lesson well. He had, of course, seen his mentor, Ernest Lawrence, build those early cyclotrons, culminating in the then-immense sixty-inch machine. But his real school in big physics was the war effort itself, where he, unusually, worked on both the radar project at MIT and the Manhattan Project in Los Alamos. They changed his work style abruptly and permanently; linework became the rule. Turning his postwar attention to an ever-escalating series of weapons projects, in the early 1950s Alvarez launched the first of the physics factories: his liquid hydrogen bubble chamber and the laboratories that surrounded it. Nothing like it could have been imagined from the halls of the Physikalisch-Technische Reichsanstalt. Here in Berkeley in the late 1950s—as at the war laboratories—there were day shifts, night shifts, and swing shifts running round-the-clock experiments. Rows of (mostly women) scanners computer encoded data from the stereoscopically projected bubble chamber images. Other operators manipulated the data on the computer, running simulations and comparing the "real" to the artifactual information, occasionally revealing a "bump" in the data that signaled the discovery of a new particle. Even theory had changed its configuration during the war: theory groups calculated furiously at problems the likes of which had never before been undertaken. What was the critical mass of uranium and plutonium needed for detonation? Calculate too low and the bomb might fizzle, wasting hundreds of millions of dollars and possibly revealing progress and even materials to the Axis; calculate too high and the physicists of Los Alamos, Hanford, and Oak Ridge might incinerate themselves and their nearest hundred thousand neighbors. Calculations were needed, too, for efficiency of the detonation and its effects, for the radar antennae, and indeed for the whole of microwave physics. For industrial as these processes were in size, no engineer from Du Pont or RCA was in a position to calculate much of anything about microwave electronics or nuclear physics. When war ended, theorists, like experimentalists, were ready to take on problems of a scale and difficulty they had assiduously avoided before the war.[4]

For our purposes let us take as emblematic of laboratory high modernism a mental picture of Einstein suspending a tiny magnetic cylinder on a fragile glass fiber in the laboratories of the Physikalisch-Technische Reichsanstalt. Representing late modernism might be Luis Alvarez, directing his "chief of staff" to survey the output of scanning shifts, programming teams, and cryogenic engineers. Einstein, at least briefly, could embody the crossing point between engineering and physics; Alvarez's group managed the relationship between them. Einstein's physics was a physics conducted alone, occasionally with a technician or a collaborator, but structured under his individual authorial name. Alvarez's name too stood for the bubble chamber projects, but as CEO of a very large corporation.

5. Postmodernism: The Hybrid and Data-Based Laboratory

If there is one place the term *postmodernism* carries a certain clarity, it is in the architectural contrast between buildings that aim for unity and homogeneity—the Seagrams building in New York City, for example—and the self-consciously mixed idiom of a structure like Philip Johnson's AT&T headquarters just a few blocks away. This architectural contrast is helpful in articulating the differences between the modern laboratory of Einstein's magnetism apparatus at the PTR or Alvarez's bubble chamber laboratory centered on his hundred-gallon vat of liquid hydrogen, and the next era of hybrid colliding-beam detectors that I will term *postmodern*. In the hybrid detector, with its many subdetectors, we have a dispersion of both knowledge and authority. The individual subdetectors clustered within the massive hybrid are built and designed separately, they are serviced separately, they are simulated separately, and ultimately they produce data separately. Indeed, they are so autonomous that it is frequently the case that the very effort to integrate hardware, software, and data becomes one of the defining features of the collaboration. Contributing to this internal quasiautonomy is the increasing pressure by participating institutes, universities, even countries, to be able to identify a clear portion of the experiment that is "theirs." Indeed, the political economy of contemporary high-energy experimentation is such that it is virtually impossible to find ten or twenty university or national groups all willing to permanently subordinate themselves to one leading institution—much less an individual.

This contrast is most vivid at the extreme, and the most extreme detectors are without any doubt the twin behemoths that were to have become the electronic nodal points of the once-but-never Superconducting Supercollider (SSC). Each of these two detectors was to have been built at a cost of some $1 billion each with teams numbering nearly a thousand Ph.D.'s each at the time of the SSC shutdown. Connected by computer links and spread over Europe, the United States, and Japan, these "collaborations of collaborations" little resemble the crew of twenty or so physicists Alvarez held under his direct command. Hanford was not "like" a factory, it *was* a factory; the SSC and the large-scale colliding-beam facilities were not "like" multinational corporations, they *were*, in all but name. And just as one hesitates before answering the question "Where is Exxon?" one would do well to pause before giving a street address for a major high-energy physics collaboration. Not only was it expected that the SSC would export data to be analyzed by any of the thousand or so participants in each "experiment," but *control* of the machine itself could be run remotely from several places in the United States, Europe, or Japan. The spokesperson may be in one place, the machine in another, the data analyzed in twenty-five different laboratories across four countries, and the collaborators are unlikely to have met each other.

How do technology and science interact in such a decentered, data-based environment? Let's take a particular and slightly earlier instance, the first big Time Projection Chamber (TPC), a detector developed for use between 1974 and 1992 at the Stanford Linear Accelerator Center (SLAC) in Northern California.

Despite five decades of intense collaboration between physicists and engineers, the lines of division remained clear, even where the two groups worked in the closest coordination. For as complex as the relation is between physicists and engineers, it surely *underestimates* the usual friction between technical subcultures; what is odd is not that such sparks flew, but that so few did. Consider, for example, the development of the TPC, a particle detector invented by the experimentalist David Nygren at Berkeley in 1975. The idea was to build a large chamber capable of electromagnetically preserving the ionized tracks left by passing particles, and to float these tracks, intact, to endplates that would use their drift time to reckon the exact geometry of the original trajectories. Within months, plans began to scale up the prototype to a massive, industrial scale.

In February 1979, Nygren and the project manager Jay Marx set out the highly heterogeneous set of constraints within which the system had to function, given the threat of a huge cost overrun.[5] For example, the TPC designers had to maintain the distribution of resources and work within "core detectors" (TPC, drift chambers, and magnets) as "sociologically viable within the collaboration." Some pieces went to Lawrence Berkeley Laboratory, others to SLAC itself; some were farmed out to the University of California-Riverside. Even the laboratory director Wolfgang Panofsky wanted to hold an identifiable (spatial) portion of the detector in reserve against cost overruns, as a kind of "security."

In short, the architectural hybridity of the detector embodied the multiplicity of institutional aspirations of the different constituent groups at work on the TPC, and it was politically useful, even necessary, for these separate components to have well-defined institutional homes. To the designers, modifying the detector meant operating within a set of constraints imposed by the administrative/accounting branch of the laboratory, the scientific demands of the particle physicists, and the "sociological" impositions of the participating laboratory entities.

Not only did the changing political economy of detector-building render it impossible to lodge a major project like this in a single place (as the Atomic Energy Commission could and did do in the Alvarez era), it was difficult for any single country, even one with a physics budget the size of that of the United States, to go it alone. Among other attempts, the Americans approached two Japanese groups. The meeting of two cultural worlds, even expressed through the technicalities of physics, was not simple.

An anthropologist of science, Sharon Traweek, recorded the simultaneous awkwardness and productiveness of cultural differences when a Japanese and American particle physics group began collaborating in Japan. To the Japanese, the Americans seemed shockingly sloppy, brutally rude to one another, and irresponsibly expressive of their views. But taking these "faults" as resources, the Japanese group began training their own students (on the Americans' model) to "forage" for spare and useful parts from discard bins, and sent the childishly petulant foreign physicists to plead with (and embarrass) the accelerators' leaders into giving the joint team more beam time (Traweek, 1992, 1988). We see a striking antiparallel in the TPC project, where the team sought to entice the Japanese into the collaboration (Galison, 1997, Ch. 7). As one Yale physicist reflected while the collaboration contemplated going to the Japanese for resources:

> On the question of how the Japanese are as collaborators, they are, from an American's point of view, ideal. They seldom have their egos involved in their work and thus do not waste time in arguing simply to defend a position. This cannot be said of many groups of people in general, and especially of physicists. They do, however, work very hard, meticulously, and accurately.
>
> If there is a fault in their work it is their lack of imagination. They seem to rely very strongly on authority and seldom deviate from prescribed procedures. For us this attribute will probably enhance our collaboration rather than detract from it (Marx, *Correspondence,* January 4, 1979).

As with the solution to the problem of the directorate, the problem of the multinational collaboration was solved spatially. By tying the Japanese contribution to a specific component (the cylindrical calorimeter), the Japanese government would have "an 'identifiable' area which is good politically" (Marx, *Correspondence,* February 5, 1979).

Each group's stereotypical vision of self and other worked to reinforce a collaborative reality, not just between different national groups, but among the myriad of distinct groupings that constituted the collaboration as a whole. It was anything but homogeneous. Indeed, when an SLAC cartoonist parodied a major detector built shortly before TPC, he sketched it six different ways: *inter alia,* as gold for the accountants, as a metal detector for the physicists, and as a disassembled watch for the mechanical engineers. Reproduced later in the SLAC in-house journal *Beam Line,* this cartoon has much to say about large-scale technical systems in the postmodernist frame. In particular, the view from on high is not the view from other perspectives; the cylindrical calorimeter becomes a spatialized political chip in Tokyo, a line-item on the third floor of SLAC's central laboratory, and a quark jet finder to the experimental physicists.

I make this point about heterogeneity *despite* the fact that experimental physicists, such as two of the collaboration leaders Dave Nygren and Jay Marx, worked constantly with engineering problems associated with the TPC. No amount of collaboration could erase the sense of difference between engineering and physics that remained a pervasive and fundamental piece of the participants' self-identification. At one point in July 1979, the volatile relations between engineers and physicists threatened to detonate the entire project, as Marx confided to Nygren in a confidential memorandum:

> The role of engineering personnel vis-à-vis physicists in reaching design specifications, design concepts and in the execution of detailed engineering design has not been satisfactory. In my view, the physicist's proper role is to push the technical design as far as he is able in the direction of maximizing the scientific output of the device being designed. His job is to demand specifications as determined by scientific goals, even if these specifications are at the "state of the art" level. The engineer, on the other hand, must be more conservative and realistic. His proper role is to advise on what can be done at what cost, in what period of time (Marx, *Correspondence,* July 17, 1979).

Marx saw the frequent changes of plans as taking a catastrophic toll on the engineers. Several had already quit the laboratory in large part because of the rapid fluctuations, and others were moving in the same direction.

Marx then continued: "The engineer's job is to inject reality into the physicist's dreams and then to help to translate their mutually acceptable specifications into a working piece of hardware" (ibid.). But in order for the engineers to act as a reality principle checking physicists' tendency to wish-fulfilling dreams (to stop their physicist colleagues from "chasing butterflies" as Greenewalt put it during the war), the engineers' culture had to be able to stand up to the coercion of the physicists. Collaboration demanded what a deep power imbalance prevented. Marx went on:

> Clearly this [injection of reality] requires dialogue between physicists and engineers. The engineers must be willing and able to engage physicists in such technical give and take. . . . Such technical discussion between engineers and physicists requires confident engineers who understand what this system demands as well as strong, confident leadership in the engineering department. Without such debate, the engineer is left with the frustrating task of designing to unreasonable specifications, and then having to recycle the design when, at some later time, the physicist himself embraces more realistic specifications (ibid.).

At the root of these struggles were deep-lying differences in technical culture. For the engineers, semi-works, design freezes, and input-output analyses were the bread and butter of their existence. What to them felt like good engineering appeared to the physicists as overly conservative know-nothingness. For the physicists, good experimentation meant the application of "state-of-the-art" devices; it meant constantly upgrading their plans to realize in circuit board and steel the best design they could produce. To the engineers, the physicists' "recycling" was childishly unrealistic, it was butterfly chasing and dreaming. Multiply these sorts of disputes twenty-fold to include the gaps between theorists and experimentalists, software and hardware workers, mechanical and electrical engineers, and all the other groupings and subcultures that make up a heterogeneous collaboration, and one has the real situation of a dispersed and hybrid collaboration. Understanding how such disunified collocations come to form true collaborations is, I would argue, the single most important practical, technical, theoretical, and even philosophical problem of late-twentieth-century, large-scale experimentation.

6. Heterogeneity without Hierarchy

Returning now to the problem of writing a history of the borderland between science and technology, it seems clear that a certain battle has been won. In many systems, ranging across a myriad of disciplines, the old view of a fundamentally unidirectional flow from science to technology has been supplanted. From inertial guidance, ammonia synthesis, black-body radiation, power generation, electrodynamics, telegraphy, thermodynamics, cyclotrons, steam engines, or renormalization, some less hierarchical picture of science and technology must be introduced. That said, where to begin?

In my judgment, we should not replace the assumption of universal hierarchy in which science dictates to technology with the assumption that there is one, homogeneous entity: "technoscience." More specifically, in the effort to dislocate the science-centered picture of the science/technology relation, it is tempting to dismiss the very real differences between the technical cultures of engineering and of physics or chemistry, as well as differences and strains within the disciplines themselves. Tensions of this sort manifest themselves in a variety of ways, between electrical and mechanical engineers, between theoretical and experimental physicists, between physicists and engineers, between physicists and chemists, between biologists and physicists.

Instead of simply forcing together the disparate cultures of the technical and scientific worlds and labeling their connections as "symbiotic," "interactive," "interpenetrating," or pointing to the "mutual exploitation of resources," the task before us is to explore the detailed nature by which the groups communicate with one another and coordinate their actions and beliefs.

For it is my suspicion that it is at these seams—these carefully chosen boundary points between cultures—that an exchange takes place of extraordinary interest. Elsewhere, I have called this area "the trading zone." It is there that the Alvarezes, Wheelers, and Comptons learn their flow-chart organizations and learn how to mesh their physicists' concerns with those of the engineers. Sometimes this trading zone has a spatial existence—as in the rooms of the MIT Radiation Laboratory where a "pure" physicist like Julian Schwinger could learn input-output analysis from the electrical engineers, literally across the table (Galison, 1995, 1996, 1997). In other sites, the exchange exists in administrative architecture, or pedagogical coordination. But wherever such exchange occurs, it provides the historian of science and technology with a nexus of joint concerns that takes us beyond the denial of difference to a complex net with a patchwork topology of knots, gaps, and overlays.

In an effort to capture this coordinated but heterogeneous set of scientific subcultures, I have been increasingly discontent with the old notion of "translation" between the different languages of science. Instead, I have been interested in what really happens at the boundary between language groups that have to work together. They establish local exchange languages: the specialized jargons, pidgins, creoles that anthropological linguists have been studying so productively. Though I have to refer to my work elsewhere for a much more extensive discussion, the idea is that when biology and chemistry meet they don't interact by instantaneously translating between the fully formed formalism and practices of their separate disciplines. Instead, they work out local, shared sets of practices and terms, a "trading language" aimed at solving problems in the borderland, not reconciling morphology and quantum chemistry. As the interdiscipline grows, the pidgin becomes a creole; that is, a language rich enough to allow someone to grow up within it. Finally, a discipline such as biochemistry is born, a new and robust technical culture where a specific set of techniques from biology is thoroughly coordinated with a specific set of techniques from chemistry.

So I end with a problem, and the schematics of a solution. How does one think about an entity such as large-scale physics, which on closer examination is constituted by a panoply of different sub-cultures? There are subcultures in which members identify themselves by their affiliation as engineers, as theorists, or as experimentalists. And upon closer examination, even these divisions split apart: mechanical engineers, electrical engineers, cryogenic engineers all view themselves as having distinct modes of analysis, training, and institutional incarnations. Similarly, theorists understand both implicitly and explicitly the differentiation between field theorists, phenomenologists, and mathematical physicists. Our job is not merely to label the mutual relations of these differing groups, but to explore the dynamics and evolving intermediate zones in which they coordinate their actions, and to understand when a dynamic hybrid creole has emerged. Here arises a difficult question. In the age of "big science," the different cultures manifest themselves in sociologically distinct entities—it is not hard to distinguish the computer simulation team from the electrical engineers within a colliding-beam experiment. But in the earlier, smaller-scale, modernist phase of physical experimentation, as with our instance of Einstein at the PTR, how does one think about the combination of the language of electrical engineering and the language of physical theory? My own inclination is that technical languages, not unlike informal languages, are never private affairs, and that even without sociological division, it still makes sense to distinguish the languages and to examine the modality of their coordination from pidgin through creole.[6]

If I have identified this problem of narrating a coordinated, but not homogenized, history with multiple subcultures as critical, it is because I believe that the problem of narrating the story of an inhomogeneous culture, with its distinct, often immiscible subcultures, each with its own history, its own goals, and its own standards is far more than a problem for technical historians of recent science. It is the central problem of writing cultural history in the 1990s.

Notes

[1] Inspired in part by Thomas Hughes's work on the systems approach to treating technological alongside scientific work (Hughes, 1983), a variety of historians and sociologists have begun to dismantle the top-down picture. Introducing their outstanding collection of essays on technological systems, Trevor Pinch and Wiebe Bijker argued that: "'Technology/science,' 'pure/applied,' 'internal/external,' and 'technical/social' are some of the dichotomies that were foreign to the integrating inventors, engineers, and managers of the system- and network-building era. To have asked problem solving inventors if they were doing science or technology probably would have brought an uncomprehending stare. . . . Instead . . . integrating managers . . . saw a seamless web" (Bijker, Hughes, and Pinch, eds., 1987, p. 10). Ed Layton, whom Pinch and Bijker cite, puts it this way: "Science and technology have become intermixed. Modern technology involves scientists who 'do' technology and technologists who function as scientists. . . . The old view that basic sciences generate all the knowledge which the technologists then apply will simply not help in understanding contemporary technology" (Layton, 1977, p. 210). For Barry Barnes, the newer science and technology studies "recognize science and technology to be on a par with each other. Both sets of practitioners creatively extend and develop their existing culture; but both also take up and exploit some part of the culture of the other. . . . They are in fact enmeshed in a symbiotic relationship" (Barnes, 1982, p. 166). Other sociologists working on recent technology attack the same dichotomy. On electric cars, for example, Michel Callon insists that, "Sociological, technoscientific, and economic analyses are permanently interwoven in a seamless web" (Bijker, Hughes, and Pinch, eds., 1987, p. 85), and Bruno Latour encapsulates the breakdown of science and technology in his single locution, "technoscience" (Latour, 1987, esp. Ch. 4, "Insiders Out").

[2] Bruce Hunt has productively inquired into the links between the practice of telegraphy and the theory of electromagnetism (Hunt, 1991). Crosbie Smith and Norton Wise's remarkable biographical study of Lord Kelvin (William Thomson) shows another Victorian physicist-industrialist for whom there famously was no hard and fast division

between the pure and the applied, but instead a single outlook, which they labeled his "industrial culture" (Smith and Wise, 1989, p. xxi; Wise, 1988). Thomson here appears not as applying abstract science to industry, but as working the other way around: from industry to science. In nineteenth-century chemistry, Timothy Lenoir has followed the complex interaction between thermodynamics and ammonia synthesis (Lenoir, 1997, Ch. 8). On precision measurement, Simon Schaffer has shown how closely tied the development of the absolute measurement of resistance was to the standardization of resistance and the international competition over which country would lead that standardization (Schaffer, 1992).

3 Galison and Assmus, 1989; for more on hygiene and the obsession with dust, see Forty, 1986, esp. Ch. 7.

4 On the postwar development of quantum electrodynamics, see Schweber, 1994.

5 These multiple constraints closely resemble, in their diversity, those faced by MacKenzie's missile builders or Latour's electric car designers.

6 Wittgenstein, of course, argued strenuously against the notion of private language. The issue here is whether that argument carries over to the case of technical languages. But the burden of proof would seem to lie with those who thought there was a fundamental distinction between technical and nontechnical language. In the past, such demarcation criteria have not succeeded.

References

Barnes, Barry, "The Science-Technology Relationship: A Model and a Query," *Social Studies of Science* 12 (1982): 166–72.

Bennett, Jim, Brain, Robert, Schaffer, Simon, Sibum, Heinz Otto, and Staley, Richard, *1900: The New Age: A Guide to the Exhibition* (Cambridge: Cambridge University Press, 1994).

Bijker, Wiebe, Hughes, Thomas, and Pinch, Trevor, eds., *The Social Construction of Technological Systems* (Cambridge, Mass.: MIT Press, 1987).

Buchwald, Jed, *From Maxwell to Microphysics* (Chicago: Chicago University Press, 1985).

Cahan, David *An Institute for an Empire* (Cambridge: Cambridge University Press, 1989).

Einstein, Albert, *Collected Papers of Albert Einstein, Volume 5: The Swiss Years,* Anna Beck, tr. (Princeton: Princeton University Press, 1995).

Flückiger, Max, *Einstein in Bern* (Bern: Paul Haupt, 1974).

Forman, Paul, "Behind Quantum Electrodynamics: National Security as Basis for Physical Research in the United States, 1940–1960," *Historical Studies in the Physical and Biological Sciences* 18 (1987): 149–229.

Forty, Adrian, *Objects of Desire* (London: Thames and Hudson, 1986).

Galison, Peter, *How Experiments End* (Chicago: Chicago University Press, 1987).

Galison, Peter, and Assmus, Alexi, "Artificial Clouds, Real Particles," in David Gooding et al., eds., *The Uses of Experiment* (Cambridge: Cambridge University Press, 1989).

Galison, Peter, and Hevly, Bruce, eds., *Big Science: The Growth of Large-Scale Research* (Stanford, Calif.: Stanford University Press, 1992).

Galison, Peter, "The Trading Zone: The Coordination of Action and Belief," presented at TECH-KNOW Workshop on Places of Knowledge, Their Technologies and Economies, UCLA Center for Cultural History of Science and Technology (March 31, 1990).

Galison, Peter, "Context and Constraints," in Jed Buchwald, ed., *Scientific Practice* (Chicago: Chicago University Press, 1995).

Galison, Peter, "Computer Simulations and the Trading Zone," in Peter Galison and David J. Stump, eds., *Disunity of Science* (Stanford: Stanford University Press, 1996).

Galison, Peter, *Image and Logic: A Material Culture of Microphysics* (Chicago: Chicago University Press, 1997).

Greenewalt, Crawford H., *Diaries,* Hagley Museum and Library Collections, courtesy of Hagley Museum and Library, Wilmington, Delaware.

Harman, Peter, *Energy, Force, and Matter* (Cambridge: Cambridge University Press, 1982).

Heilbron, John, Seidel, Robert W., and Wheaton, Bruce R., *Lawrence and his Laboratory: Nuclear Science at Berkeley, 1931–1961* (Berkeley: OHST, University of California Berkeley, 1981).

Hounshell, David, and Smith, John Kenly Jr., *Science and Corporate Strategy: Du Pont R&D, 1902–1980* (New York: Cambridge University Press, 1988).

Hounshell, David, "Du Pont and the Management of Large-Scale R & D," in Galison, Peter, and Hevly, Bruce, eds., *Big Science: The Growth of Large-Scale Research* (Stanford, Calif.: Stanford University Press, 1992).

Hughes, Thomas, *Networks of Power* (Baltimore: Johns Hopkins Press, 1983).

Hughes, Thomas, "Einstein, Inventors, and Invention," *Science in Context* 6 (1993): 25–52.

Hunt, Bruce, *The Maxwellians* (Ithaca, N.Y.: Cornell University Press, 1991).

Jungnickel, Christa, and McCormmach, Russell, *Intellectual Mastery of Nature*, vol. II (Chicago: Chicago University Press, 1986).

Kevles, Daniel Jerome, *The Physicists: The History of a Scientific Community in Modern America*, 4th ed. (Cambridge, Mass.: Harvard University Press, 1995).

Latour, Bruno, *Science in Action* (Cambridge, Mass.: Harvard University Press, 1987).

Layton, Ed, "Conditions of Technological Development," in *Science, Technology and Society: A Cross-Disciplinary Perspective*, I. Piegel-Rösing and D. de Solla Price, eds. (London and Beverly Hills: Sage, 1977), pp. 197–222.

Lenoir, Timothy, *Instituting Science: The Cultural Production of Scientific Disciplines* (Stanford, Calif.: Stanford University Press, 1997).

Leslie, Stuart W., and Hevly, B., "Steeple Building at Stanford: Electrical Engineering, Physics, and Microwave Research," *Proceedings of the IEEE* 73 (1985): 1169–80.

Marx, Jay, *Correspondence*, Jay Marx Papers, Time Projection Chamber Research and Development Administrative Records, 1972–83, Archives and Records, Lawrence Berkeley Laboratory, Berkeley, California.

Meyer, Hannes, "The New World," *Das Werk* (1926), typescript English translation from Meyer Papers, Frankfurt.

Needell, Allan, "Nuclear Reactors and the Founding of Brookhaven National Laboratory," *Historical Studies in the Physical Sciences* 14 (1983): 93–122.

Pyenson, Lewis, *The Young Einstein* (Bristol: Hilger, 1985).

Sanger, S. L., *Hanford and the Bomb* (Seattle: Living History Press, 1989).

Schaffer, Simon, "A Manufactory of Ohms," in Robert Bud and Susan E. Cozzens, eds., *Invisible Connections: Instruments, Institutions, and Science* (Bellingham, Washington: SPIE Optical Engineering Press, 1992), pp. 23–49.

Schweber, S.S., "The Mutual Embrace of Science and the Military ONR: The Growth of Physics in the United States after World War II," in *Science, Technology, and the Military*, E. Mendelsohn, M.R. Smith, and P. Weingart, eds., *Sociology of the Sciences*, vol. I (Dordrecht: Kluwer, 1988), pp. 3–45.

Schweber, Silvan S., *QED and the Men Who Made It* (Princeton: Princeton University Press, 1994).

Segrè, Emilio, *Enrico Fermi Physicist* (Chicago: University of Chicago Press, 1970).

Smith, Crosbie, and Wise, Norton, *Energy and Empire* (Cambridge: Cambridge University Press, 1989).

Traweek, Sharon, *Beamtimes and Lifetimes* (Cambridge, Mass.: Harvard University Press, 1988).

Traweek, Sharon, "High-Energy Physics in Japan," in Galison, Peter, and Hevly, Bruce, eds., *Big Science: The Growth of Large-Scale Research* (Stanford, Calif.: Stanford University Press, 1992).

Wise, Norton, "Mediating Machines," *Science in Context* 2 (1988): 77–114.

Science and Technology: Biology and Biotechnology

BY JOSHUA LEDERBERG

I HAVE understood *science* to mean our effort to comprehend ourselves and the world we inhabit and to try to make or impose some sense on it. When Einstein winced at the thought that God might play dice, he was reflecting a deep unease about senselessness, not that he would be troubled about building a machine.

Technology is the use of scientific knowledge and insight to make things or practice processes that are inspired by someone's practical advantage.

Scientists cannot help stumbling onto technical innovations, and often justify investment in their work by the promise that innovations will follow. And technical praxis may be the means of corroborating the objective utility of a scientific generalization, carrying it beyond the domain of social construction of "the truth." On the other hand, praxis often uncovers limitations in scientific understanding and brings forth phenomena that provoke profound inquiry.

When a scientific principle is studied in the laboratory, it is subjected to controlled trials, the essence of which is to limit the number of incident variables. In technical or clinical application, or in natural history, nature—not human judgement—brings into play new variables, including some that had not previously been adjudged to be relevant. Thus, the discovery in 1944 that the genetic material was composed of DNA, took place at the Rockefeller Institute as the byproduct of studies on the classification of bacteria causing pneumonia.

From the very beginning, it was clear to the emerging practitioners of molecular biology—and I have been at it since 1944 or earlier—that there would eventually be enormous practical fallout from these axially important findings about the gene and the cell. In fact, as the years went by, I would lament that it took close to thirty-five years before it could be said that anyone's life had been saved by our knowledge of the structure of DNA. Until well into the 1960's, genetics was a marginal discipline in the teaching of medicine—having founded the departments at Wisconsin (1955) and Stanford (1959), I can testify to the struggle. Today, pharmaceutical, immunological, and pathological science and technology are dominated by the iconic vision of the double helix. We could not have begun to comprehend what the human immunodeficiency virus (HIV) was without that; nor could any of the now burgeoning drug treatments have been developed.

For now, I will rely on an article (Lederberg, 1993) that appeared originally in a series on molecular medicine in the *Journal of the American Medical Association* to spell out detailed examples. Besides medical application, DNA analysis has furnished the most spectacular advance in forensics, for criminal identification and the labeling of human remains, as well as the authentication of paternity. And biotechnology is beginning to make a dent in agriculture and in a few industrial chemical processes. For the latter there is much impetus from the avoidance of nonbiodegradable solvents, and from the positive use of biotechnology in environmental cleanup. But we are not yet past orchestrated fear of "genetically manipulated foods," especially in Europe, where it may of course serve as a nontariff barrier to trade.

The biotechnology industry had a market capitalization of $52 billion (on current sales of $9.3 billion, and research and development expenses of the same order) (Ernst and Young, 1996). Even today's skeptical market still has some optimism for future prospects. The picture is being blurred by the consolidation of many smaller biotechnology companies into the pharmaceutical giants, and by the belated incorporation of DNA-based strategies into their own research doctrine. Thirty-five years ago, I had zero takers when I tried to interest the pharmaceutical industry in a combinatorial (Darwinian) approach for drug discovery. Today, this is described as the central paradigm. My wise friends have told me that the resistance came from the establishment chemists who would be pained ever to let a compound out of their synthetic laboratories until they had purified and verified its structure. Anything else would be "Schmer chemistry," epitomized by Dr. Gottlieb's invective in *Arrowsmith*. It was bolstered by the expectation that theoretical structural analysis and x-ray determinations of drug-receptor fit would provide a rational basis for drug development. Only then could the pharmacologists and toxicologists be given access to it. My rejoinder had been that life could never have started on Earth if the Ur-Chemist had been similarly constrained.

Expectations of social utility and of profit are now plainly far more powerful motives for pursuing biological science than when I entered that vocation a half-century ago. Robert Merton's characterizations of scientific norms, may I call them the dignity of my profession, are a familiar portrait of what I remember. Scientists have always been jealous of their prestige, or priority in discovery; today many of them may have far more material pressures on their interpersonal and moral behavior. By its impact on technology, science may also matter more to the social body, with promises of medical advance and threats from earth-consuming pollution and weapons. The "rest of culture" does not notice, both in large (but perhaps now becoming asymptotic) governmental support for research, and in widely voiced anxieties about being overrun by technology that burgeons faster than anyone can understand its full implications.

Technology does, of course, guide the possibility of investigation in the modern laboratory, where string and sealing wax may be hard to find. Biology was one of the last of the natural sciences to eschew heavy-metal technology. I began my career, and as far as feasible still try, to practice science where the weight of ideas outbalances that of the equipment. (Not always: some of my experiments were carried out on Mars, thanks to rather large space rockets and a cast of thousands at the NASA command centers and engineering development programs. But those rockets would have been built and paid for regardless, and put to even more problematical applications, if exobiology had not been on the table.)

But coming back to earth, biology would be a poor competitor with physics in an Aristophanean competition weighing chariots against fleets of ships in the literary competition of Aeschylus versus Euripides.

Stacked up against the major accelerators, our moderate size machines are represented by six-digit, not ten-digit, investments: the electron microscope, x-ray diffraction, Nuclear Magnetic Resonance, and the robotocized gene-sequencers and -synthesizers; and the latter can be commercialized, retailed, and leased out for a few hundred dollars a shot. Most laboratories budget more for graduate and postdoctorate assistants than for equipment. In fact, some of the most rewarding technical developments have eventuated in becoming smaller and cheaper. The simple agar gel electro-powered diffusion (gel electrophoresis) has replaced expensive ultra-centrifuges. And the polymerase chain reaction (PCR) technology was named "molecule" of the year for having democratized access to DNA. The kits for PCR cost just a few hundred dollars and they offer the detectability of just single molecules of DNA and easy clues to the structure of the tiniest samples.

In fact, we have an agenda for productive biological research for the next century that requires no new ideas at all: to tease out the one hundred thousand or so genes that populate the entire human genome (which has room for ten million). This can clearly be done within the scale of research and development investment of the biotechnology industry, and some multiple to spare for the real task of teasing out the functions and the interactions of all these genes. Just within the last year we have seen the mapping of several bacteria and of yeast; and the first order of maps of the human are in sight. Some firms have the expectation that by this mechanical sequencing of a gene, or tag ends of it, they may gain property rights to any further use of that knowledge—that still has to be fought out in the courts and in the Congress.

The mechanical production of all this new knowledge is quite wonderful; but how widely is it understood that it is just the first step? I have no doubt at all that some significant percentage, perhaps ten thousand genes, will prove to have important biomedical application. But history tells us that to bring any one to practice involves bare minimum investment of $100 million each. So we do have a triage problem of the scale of the GNP; that is, measured in trillions of dollars. The effort, even when trimmed, might exhaust all of our intellectual as well as financial resources; and in that setting will we again have fresh breakthroughs like those of 1944? The opening up of fresh paradigms, beyond what can be programmed from Washington, or even by a hidden hand that is stuffed with lucre?

References

Ernst and Young, *Biotech '96* (Palo Alto, Calif.: Ernst and Young, 1996).
Lederberg, Joshua, "What the Double Helix (1953) Has Meant for Basic Biomedical Science: A Personal Commentary," *Journal of the American Medical Association* 269:15 (1993): 1981–85.

PART 5

Political Life

What is the relationship between our means of communication and our sense of community? How may changes in the ways in which we communicate challenge our system of democracy and our system of free expression? How have these changes been reflected in our laws?

Introduction

BY IRA KATZNELSON

This section on political life is broadly devoted to a considera-tion of how shifts in the means and terms of communication affect democracy, free expression, and the law. Alan Ryan revisits old themes in liberal discourse and theory, especially fears about the limited capacities and character of mass opinion. He worries thoughtfully about how technology is misshapen by commercial pressures and decisions about its use. The result is a paradox of less meaningful information in the context of more. His anxiety about late-twentieth-century society's capacity to govern itself in the face of these threats to genuine publics is complemented by Paul Gewirtz's learned disquisition on the mutual dependence of technology and law in our time. Faced with new technologies, American courts both adapt old principles and create new ones. The hybrid they produce has a profound effect on just the Deweyite issues Ryan raises so forcefully.

By now, students of the media have taught us that its capacity to affect behavior and belief can be overstated. Citizens are not just passive recipients of mass-produced messages. Equally, however, it is hard to imagine how a democracy can effectively function with-out means to transform audiences into publics. If this was the task Dewey sought to convene at our century's start, it is ever more pressing today.

Read together, Gewirtz's and Ryan's papers prod those of us concerned for the fate of liberal democracy to think hard about how the utilization of the new communications technologies and the legal rules that alter the probabilities of their application are challenges that in part are amenable to informed political and legal control. Not just our courts, but our politicians and political class, as well as ordinary citizens, have momentous decisions to take, albeit under conditions of considerable uncertainty.

Because these choices demand informed debate, the very process of convoking discussion can contribute to an alleviation of the worst case scenario Ryan poses, and, in turn, can shape the climate within which key court cases of the kind Gewirtz discusses will be decided. Alas, judging from these articles, there is precious little evidence that either the current legal regime regulating broadcasting and the media or the judgments made by its managers are enhancing the public conversation we so badly need.

Exaggerated Hopes and Baseless Fears

BY ALAN RYAN

I.

THIS article investigates thoughts and themes that I have encountered recently concerning technology and culture, focusing most heavily on two ideas that are particularly contentious. The first is the thought that the communicative possibilities opened up by the Internet and other novel forms of communication may create genuine, if "virtual," communities. I am dubious about the plausibility of this, and take a few paragraphs below to say why. The second issue is connected, but not identical, and concerns the way individuals respond to the mixture of personal and impersonal communication possible through e-mail and the like. Again, the thought that a person might acquire a new "virtual identity" under such conditions strikes me as implausible, but I shall give it a run for its money here. In general, I think that there is a temptation to elevate the social impact of almost every new communicative technology to a philosophical plane on which it does not belong. The sociology of these matters is interesting enough in its own right not to need turning it into metaphysics.

The burden of this essay is, as the title suggests, that neither the greatest hopes nor the worst fears have been realized of those who have wondered about the connection between the technology of communication and its content on the one hand and its political and moral impact on the other. One thought, and it might be taken as either hopeful or fearful according to taste, is that the most recent technological shift—that is, the arrival of the Internet—has enabled the creation of "virtual communities," which is to say groups of people who have no direct, face-to-face, physical contact, do not have what one might call unmediated mouth-to-ear communication, but who nonetheless share such a community of common interests and concerns that they constitute a true community. The fearful interpretation is that such virtual communities are sufficiently like traditional communities to distract our attention from "real life," and therefore to impoverish our normal, natural, and real social interaction. The cheerful interpretation is that people who for whatever reason have a thin time of it with finding friends to talk to directly about matters in which they are interested may be able to find companions at a distance and via their computers. The obvious candidates for such a blessing are academics who live and teach in particularly remote parts of the country—philosophers teaching in remote parts of North Dakota, for instance, whose nearest intellectual companions are three hundred miles away in Canada.

The effects of this particular mode of communication have some curious features, to which I shall return shortly. But the pleasures of being able to discuss matters of common interest with like-minded persons who cannot be actually present can hardly be denied; the number of discussion lists in both the commoner and the more arcane reaches of philosophy, for instance, is testimony to that. The question to which I should turn, however, is whether what is thus created is a community in the sense in which political theorists have been interested. The answer cannot rest on an appeal to the plain meaning of the word *community*. Just what constitutes a community is a political question, not a matter of looking up the answer in a dictionary. The point by which I am moved is this: the idea of *community* is, as the term suggests, so loose that any sort of common concern may in principle give rise to the locution of the "x-ing community," where *x* can be almost anything you care to think of—the knitting community or the snorkeling community as readily as the Heidegger-reading community. Think of an activity that people might have a common interest in pursuing, and they can plausibly describe themselves as a community; even "the self-ostracizing community" sounds plausible until the paradox sinks in. The argument therefore must be that the face-to-face community emphasized in the sociological tradition possesses a political priority rather than an epistemological one.

That is to say not much more than what I say at the end of this essay: the participatory concern with freedom of speech is a concern that people should be able to communicate what is on their minds when they have occasion to gather, and especially when they have occasion to gather for a political reason. One may draw the net of the political pretty generously, including in it such things as gathering in a union hall to elect officials or to ballot on a strike, or gathering in a school room to organize a petition for better teaching or to overturn an ordinance allowing creationist superstitions to be passed off as science. The crucial element is the element emphasized in Robert Putnam's excessively quoted article, "Bowling Alone" (1995): people have to do it together, not just in the same place, but as organized, or semiorganized, face-to-face groups. What is it that such groups have that makes them the basis for a real rather than an ersatz sense of community? I think that it is no one thing in particular. Sometimes, it is that they provide a sort of common discipline—it is harder to talk complete rubbish in front of people with whom one must cooperate day after day, just as it is harder to lie to them, too. Sometimes it is that they demand a contribution of time and effort and intelligence to sustain them, so that they provide a sort of political education and act as training in the skills of persuasion and organization. Of course, all this is *ceteris paribus*—some people talk nonsense to their friends for thirty years without noticing, and some people are blatant free-riders under all conditions, and some people learn nothing except how to bully others when they hold minor office. But one way or another, this interaction usually has such effects, as everyone from Aristotle through Tocqueville and John Stuart Mill has observed. To say this is not to deny that the virtual community may do very good things for lonely philosophers; it is only to deny that these things have the political benefits that I am concerned with here. Most people would like to know if there is sentient life elsewhere in the galaxy; it would tell us that we are not alone. Nonetheless, the knowledge that we were not alone would be far from establishing a galactic community. The Internet is good at reassuring people that they are not alone, and not much good at creating a political community out of the fragmented people that we have become.

Even at the level of doing good things for individuals who discover that they have friends in cyberspace, it is a mixed phenomenon. The thought that e-mail chatter and postings to lists allow people to explore their personalities and perhaps acquire new identities is largely misguided. To begin with the banal: very few lists survive more than a few weeks before persons who are doubtless perfectly good-natured in everyday life become obnoxious bores. They conceive that they have been insulted and insult everyone else; they believe they have discovered the Holy Grail in some branch of philosophy and insist on describing it at length. They threaten lawsuits against their critics, whine about their malformed careers, and generally carry on in ways they haven't been permitted since kindergarten. The more elaborate suggestion is sometimes floated that people who enter chat rooms and pass themselves off as somebody else entirely are somehow acquiring a new identity; but if I pass myself off as a Chinese drag queen of uncertain age, I do not become any such thing, any more than I would do so if I played such a part in a play. No doubt some people find it an aid to the imagination to pretend to others to be what they are not, but so what? For the most part, disembodied communication where nobody has to say who they actually are results in exchanges that are vapid and repetitive just because they lack the constraints of real life. Free speech is free when it is responsible—not in the sense of being dreary and commonplace, but in the sense of the utterer having to live with the consequences of their utterances. So, I conclude somewhat gloomily that inventing a superior multiperson teletype system with added pictures is not actually the creation of the body electric or the community electric. But I more happily admit that just as the telephone did good things for many lonely people while perhaps discouraging the habit of dropping in on our friends as much as we should, so the Internet, e-mail, and the like have done something of the same. Now to my main themes.

II.

This is an essay on familiar themes. Some of what I have to say is not very novel, but I offer it as a minimal setting for my more novel remarks; the questions that these raise are interesting ones. I broach them in this opening, and say more about them at the end of the essay, which thus forms an inside-out sandwich, with the most appetizing elements on the outside framing the substance that holds it together. In essence, I claim that for the past two centuries, both the hopes and fears of most social theorists and commentators have been pinned on the mass qualities of the mass communications media; only rather rarely has there been much fear of their pluralistic (or at any rate plural) character. Today, however, it is their fragmented and plural character that may cause us the most anxiety, partly for reasons that occurred to some unorthodox conservatives and their unorthodox radical allies in the nineteenth century, as well as for reasons peculiar to ourselves.

Let me sketch a few of the hopes and fears of my title to illustrate what I mean. In the early-nineteenth century, the rise of popular newspapers, cheap fiction, and instructive pamphlet literature of the sort put out by the Society for the Propagation of Useful Knowledge—memorably mocked by Thomas Love Peacock as "the steam intellect society"—conjured up a variety of fears. One was that high culture and discriminating taste would be washed away by a middlebrow culture and by the cult of merely useful information. It is outside the scope of this essay to say much about these cultural anxieties, other than to remind you that they launched the English tradition chronicled in Raymond Williams's *Culture and Society* (1958), and also prefigured the anxieties of many 1930s and postwar American intellectuals. I focus here on the more narrowly political argument. It is this that will bring me back to my theme at the end of this essay. This is the lament, to be found in Carlyle, Coleridge, and in the twenty-five-year-old John Stuart Mill, that a mass civilization was devoid of spiritual ambition. In being devoid of spiritual ambition, it was devoid of standards of excellence; the political consequence was that it could not produce political leaders (or spiritual and cultural ones, of course), and if such leaders had miraculously arisen, it would have been of no avail, since the society as a whole did not know how to follow its spiritual, cultural, and political leaders. This was not because the population at large had a fixed desire not to be governed, not to be instructed, and not to be led; it had no fixed opinions and no fixed standards. Mass opinion, on this view, was first cousin to no opinion. The larger deficiency of this condition was spiritual; the inhabitants—it should be recalled that these were the middle-class inhabitants, not the illiterate and impoverished underclass—of prosperous, industrializing countries such as Britain and the United States lacked the deep convictions about the human condition that made life intelligible and therefore tolerable. Since they had so little idea where they wished to go, they could not find a leader or leaders to guide them on their way. One might ask why that mattered; the United States had tried to create a constitution that would be a "machine that went of itself"; 1830s Britain was in the throes of a constitutional upheaval, and perhaps the English, too, should look for a machine that would go of itself and forget about individual leadership. Looking forward from the 1830s to what became Carlyle's increasingly wild enthusiasm for heroes, one might think the whole subject of political leadership better left unbroached. The answer was one that a surprising number of 1830s commentators found entirely convincing. One part of it, and a proposition that had metaphysical as well as sociological resonances, was that a belief in mechanism spreads, in a corrupting fashion, from one sphere to another. Men are not machines, and societies are not machines, but it is all too easy to fall under the sway of the economists and calculators that Edmund Burke had denounced, and persuade ourselves that we and they are exactly that. "We are grown mechanical in heart and head," said Carlyle in *Signs of the Times*, and it was a thought that resonated with Ruskin and Pugin and later with aesthetically minded British socialists. A living person is only an ambulant corpse if his life is not animated by a spirit that is distinctively his; a society is dead if it is not permeated by a common purpose. The difficulty of visualizing a form of government consistent with modern industrial societies that could satisfy the demands of such an ideal perhaps explains why so many writers in this tradition were "antitechnological," in their orientation, and to a large extent antipolitical. Even the Marxist William Morris, much later in the century, provided in *News from Nowhere* (1891) a vision of a rural society, where factories driven by a mysterious "force" were hidden away, and the only visible signs of commerce were barges full of barrels of claret making their way up the Thames. Morris's socialists lived in richly textured, articulate, and communicative societies, but they were anarchists and individualists who preferred to know rather little of the management practices of utopia.

The second response, which is much more directly to our purpose, is that leadership is necessary if society is not to be overtaken by ill-understood events. Neither Mill nor Carlyle flinched from revolution in the way we do today. Even at its most violent, the French Revolution of 1789–94 was not an exercise in systematic murder on the scale of the Soviet and Nazi death camps, and the fact that the seeds of later atrocities had indeed been sown by the Terror was hardly something that Carlyle or Mill could have understood. What they did notice was that the French Revolution occurred because of an ideological hollowing-out of the old regime. It had become what Carlyle nicely called a "sham." It looked fine and elegant, but the crowd that stormed the Bastille had only to give it one good kick and it crumbled to dust. The thought here is that societies need an animating spirit in which most of the population share—with very different degrees of articulateness no doubt—if they are to sustain themselves. A public story, an ideology, that has no grip on the imagination as well as the intellect of the people is no good.

The thought that animated critics of this persuasion was that society could only be governed in the proper sense of the word, that is, could be led in a coherent direction, if there were a coherent public opinion on which political leadership could draw. The thought was a simple one. The rise of mass communications, by which they meant essentially cheap newspapers and journals of opinion, was both the cause of and a symptom of the fact that governments now rested on public opinion, even where the constitutional arrangements were far from democratic in a twentieth-century sense. Of course, in the United States this process had even in the 1830s gone much further than in Europe, but nobody doubted that in this respect the more developed society showed to the less developed the image of its future. Not only could governments not govern against a decided public opinion; they could not govern adequately in the absence of a settled public opinion behind the measures they proposed. What was there instead of an adequate public opinion? It is stretching things to suggest that Mill or Carlyle, or even Coleridge, had a clear account of it, let alone to suggest that they had a clear account of society as divided into mutually uncomprehending groups, each with its own "public" opinion. Rather, they were impressed by the fickleness and instability of public opinion, by the uncertainty of moral and political conviction displayed by most people, and therefore by the phenomenon of what I described as "no opinion." Although they lacked the psychological theory that sustained Walter Lippmann's dissection of the way "pictures in the head" dominated political discussion, their view of the bewildered masses clutching at one simple image after another was a precursor of *Public Opinion*'s (1927) jaundiced view of things. Behind all this there was, of course, a theory, and an account of something better. This was, in Coleridge's case, the ideal of a "clerisy," an educated class that could provide moral and spiritual leadership for the society; it was an ideal that Mill latched onto in the early 1830s, though I suspect that he subsequently came to think that this early formulation was more dangerous than useful. Coleridge's wish was for a class that, at one level, combined the functions of the village priest and the village schoolmaster and thus provided coherence for the humbler life and at another level united the priest, the philosopher, and the poet in holding up an ideal image of itself to society. Mill was rather more briskly scientific minded, but essentially what he thought was missing was a class with intellectual authority; that class must be of one mind, even though it was to be of a mind made up as the result of a critical, tough-minded scrutiny of the alternatives. However much Mill admired Coleridge, he was sure that he was "an arrant driveller" in matters of economics.

Let me turn now to the more familiar tradition that saw in the rise of mass communications matter for fear or elation according to the different projects that commentators had in mind. On the optimistic side, one finds writers such as Marx, who took it for granted that every advance in communication advanced the prospects of the revolution, whether this was the development of the railways and literal communication or the invention of the steam press and literary communication. Marx was an example of the rationalist who was simultaneously a romantic. Once the workers fully understood their situation, they must surely rise in revolt against the capitalist system that so exploited them; the facts were visible, and only needed to be "internalized" so to speak for the proletariat to rise in revolt. The romantic side of this thought was that the linkage between understanding and action was so tight that the task of the revolutionary activist was only to unveil the bourgeois mystifications that might get in the way of a proper understanding. That this should require all three volumes of *Das Kapital* might be thought to be a back-handed compliment to the intelligence of bourgeois mystification. The point to notice, however, is that what to Mill or Tocqueville is something to be dreaded and avoided at all costs—that is to say, the creation of a mass opinion on political and economic matters—is to Marx the precondition of revolutionary change. What Marx feared was, on the other hand, a plurality of competing opinions that would reduce the effectiveness of proletarian organization. This presupposed that politics had been brought to a simplified binary condition, a war between the possessing classes and the expropriated proletariat. After the revolution, there would no doubt be room for a plurality of views about nearly every imaginable subject, but until then there was room only for the truth that unmasked the horrors of bourgeois society; on the other side was not permissible or even useful disagreement, but error. This was the view, expressed with characteristic starkness, of Lenin's *What Is To Be Done?* Science had no room for "free speech," and politics needed no more room for free speech than did science. Chemists did not ask for the freedom to put forward erroneous and exploded theories, and in politics those who were in possession of the truth need not be scrupulous about the rights of the deluded.

What the transition from Marx to Lenin suggests is that there was a tension in Marx's ideas between what one might call the instrumental view that effectiveness requires cohesion and cohesion requires uniformity of conviction, and the less instrumental view that there is one truth and one only, and that when that truth is understood there will naturally be agreement on it by all those whose interests do not lie in being deluded or in spreading delusion. On either view, the "mass" quality of opinion was something to be welcomed rather than feared—at any rate so long as the masses had been informed by the correct vision of things. It is not a manipulative view; it is, if one may so put it, relatively innocent about the possibilities for deliberate deception by the political authorities, and not very ambitious about the possibilities of creating a fictitious world-picture that the masses can somehow be induced to take for reality. In that sense, Marx at any rate—and, until Gramsci in the 1920s, almost all Marxists after him—did not need to confront the possibility that mass opinion might in its nature be inimical both to rationality and to truth, and might in addition be fundamentally conservative.

The thought that mass opinion was conservative by nature is one that Marx almost entertained when universal suffrage produced the election of Louis Bonaparte as prince-president and then as the Emperor Napoleon III, but it was not one that he could have done much with. But in the hands of writers such as Walter Bagehot and Henry Sumner Maine, it became part of the arsenal of liberal, antidemocratic thought. Bagehot's *Physics and Politics* (1872) is much less well-known than his essay on the English Constitution, and it now reads rather quaintly; but it is an interesting piece of work. Bagehot knew all about politics as the manipulation of opinion; it was the politics in which he was most interested, and as the editor of the *Economist* and the man who was widely regarded as a sort of unofficial chancellor of the exchequer in the 1850s and 1860s, he knew just how much opinion mattered, both to low politics and to high politics. The kind of free speech in which he was interested was not exactly that which the First Amendment of the American Constitution protects—neither on the eighteenth- and nineteenth-century understanding of the need to protect political speech against governments eager to use the law of seditious libel against it, nor the twentieth-century understanding of the need to protect sexual and artistic self-expression against the armed forces of respectability; what he minded about was the ability of the politically active to think new thoughts, to create new institutional arrangements, to conspire together for good public purposes, and to argue with one another when it was necessary to decide matters of high principle. The stretch from that to the rise of modern forms of technology looks on the face of it to be a long one. In Bagehot's mind it was shorter than one might imagine. As the title of his little book suggests, what Bagehot had in mind was that political societies evolved from simpler forms to more complex forms. As they became more complex, they ceased to live by custom; what he called "the cake of custom" had to be broken. And it was. Modern society was governed by discussion rather than by habit. But this was threatened by the rise of democracy. Democracy in turn was the product of technological change. Bagehot did not have an elaborate account of any of this, and I do not think he had a very complicated view of it all—his interests lay in the political activities of people like Sir Robert Peel, Lord Palmerston, and others of that ilk. But he saw that a combination of political pressure from the middle class, the growth of mass communications—he was, after all, the editor of a new kind of journal—and the inexorable growth of industry at the expense of agriculture was bringing with it a gradual shift towards a democratic politics. The fear was that a more democratic polity would be a less mobile, less agile, less intelligent polity. The individual self-reliance on which the mid-century liberal politics depended would not appeal to a mass public. This was argued much more fiercely by Henry Maine, whose *Popular Government* (1886) comes quite close to hysteria in its imagining of the ruin that will befall a democratic society. Still, the interest of this view lies not in the passion with which it was urged but in the fact that it was urged in defense of an essentially liberal conception of politics. Maine is well known as the author of the slogan "from status to contract" as a thumbnail description of the route taken by modernization; one way of reading *Popular Government* is that it encapsulated Maine's fear that under democracy the slogan would have to be extended with "and back again."

At all events, the pattern of the argument is not too hard to see, and its connection with technological change can be unpacked a little. At the base lie two sorts of technological change, one in the means of getting a livelihood in general, the other in the technology of mass communication, including not only cheap printing but also the rise of the railway system, together with the creation of large urban agglomerations within which travel was not only a matter of moving over short distances but also being able to use increasingly "mass" forms of public transport. On top of this base, and partially explicable in terms of it, is the process whereby people whose "opinion" was previously taken for granted, or ignored as something which had no business appearing in public, found a voice. This might have been because the lower classes increasingly found themselves in close contact with their political masters, or because their own leaders became more effective, or because they were able to come into the political community on the coattails of middle-class agitation over such issues as the repeal of the corn laws. The effect was that speech became both freer and less free; freer to the extent that new voices could make themselves heard and that governments gradually desisted in their attempts to suppress popular newspapers by means of oppressive duties and taxes on such publications, and less free to the extent that there was less imagination, less diversity, and less real political intelligence about. The present taste for talking of "deliberative democracy" is something that nineteenth-century English liberals would have found hard to understand; they thought that deliberative government was almost inevitably in competition with democratic government, and that the task of an intelligent liberalism was to balance the one against the other. This is not exactly an unknown thought in the American political tradition; it animated Madison's wish to ensure that the popular voice was heard in the corridors of power only after an appropriate filtering process had occurred.

The twentieth century inherited the nineteenth century's anxieties—and the nineteenth century's sociology along with them. Although it is a commonplace that we are a century that lost its predecessors' faith in reason, the notion that modern techniques of communication might do any number of alarming things to political debate did not wait for the World War I. Pareto, for one, and Freud, for another, held that opinion was formed by the psychological needs of those who received "information" about the world as much as, or rather than, by the logical or evidential

qualities of what they received. Freud was something of an epistemological aristocrat, therefore—only a rather tough and self-
aware person could face the evidence unflinchingly. This
"irrationalist" strain in social analysis was certainly reinforced by
the European experience of World War I, but it was only reinforced. The war provided a striking example of people acting
against their own interests, and doing so on the basis of beliefs
that were not only absurd in themselves but ones they would not
entertain when not emotionally wound up to hold them; but it
was evidence of a general capacity for self-destructive self-decep-
tion. The idea that it is the growth of modern technology, and the
technology of communication in particular, that is decisive in
forming opinion surfaced intermittently, but (and to my mind
quite rightly) it never became the leading view. The impact of
technology was essentially indirect; thus, the gullibility of people
in crowds was a theme of Roman satirists and Greek political theorists. What modern technologies of commerce and industry did
was to assemble people in crowds, where the hucksters and advertisers could get at them; they did not introduce a totally new relationship between the audience and those who manipulated it. If
one thinks of that wonderfully chilling little book, Lippmann's
Public Opinion (1927), its theme is not that newspapers mold people's opinions in sinister ways so much as that the public has no
opinions by which any rational person could try to steer a government. The reason is not itself technological; indeed, to the
extent that technology comes into the matter it is as a failure.
What people lack is a detailed, intelligent, realistic appreciation
of the information needed to run a complicated modern society
and to conduct its dealings with other such societies. In fact, given
the way Lippmann posed the problem, it was neither posed by
technology nor susceptible to a technical fix. Lippmann concentrated on the fact that a reader could spend only an average of
eleven minutes a day reading the newspaper, and then on the fact
that what was in the paper had already been distilled out of enormously long telegrams, losing in the process much of its original

connection with the world as it actually was. But, nothing short of our being everywhere present as intelligent observers could get round this fact; and even intelligent observers can disagree pretty violently about what in fact they had observed. Being able to take a film camera or sound recorder or video recorder to events and to stick them in the face of reality would not solve the problem that Lippmann identified. It was not caused by technology and could not be cured by technology. Indeed, it seems on any obvious interpretation not to be something one could cure at all. "Pictures in the head" is what Lippmann thought we are governed by, and it's hard to see how it could be otherwise, or rather, it's hard to see how one could have what Lippmann seemed to want instead.

Lippmann thought democracy meant government that was answerable to public opinion, and if there were no real public opinion it was hard to see how there could be democracy. This was why John Dewey rose so earnestly and repeatedly to Lippmann's challenge. The solution to the puzzle Lippmann posed for himself was to institute government by committees of experts who could be presumed to know what was going on and what might be done about it. In some areas, this is evidently a sensible idea; it is highly plausible that a commission of experts would do a better job of health care reform, welfare reform, and pension reform than Congress or the White House is likely to do today. But Lippmann's commissions were not to be executive, so much as a body of wise men who would vet proposed policies, point out their problems, certify the respectable, and sneer at the hopeless. It was, however, not really a solution to a problem that needed solving. That is, it has been obvious since Plato that if democracy required each citizen to have an informed and intelligent view on the details of policy, democracy is dead from the outset. To be sure, everyone is capable of having informed and intelligent views about policies: if people are fit to sit on a jury they are certainly fit to decide on the acceptability of such things as sewerage plans in their locality. Indeed, one could perfectly well imagine a system of government that would employ entirely untutored citizens in place of members of Congress, and allow them to vote yea and nay on the acceptability of policies proposed by public servants given a remit to sort out the various problems of the day. Juries do not have to be experts in crime-scene management to decide whether the police are credible; and neither the citizenry nor Congress need to be experts in the invention and building of high-tech weaponry to form a rational judgement about the credibility of the Pentagon's demands and the quality of its management. Or, to put it differently, Lippmann despaired of democracy because he rightly saw that the public—and for that matter 90 percent of the elite, too—could not perform a task that he wrongly thought democracy required them to perform.

But this is by way of preamble to my more anxious thought. Dewey responded to Lippmann in the 1920s in *The Public and Its Problems* (1927). This was a work almost wholly devoted to the problem of free speech in a somewhat "Deweyan" sense. Dewey had earlier defined democracy as a process of communication: "full and deep communication on a basis of freedom and equality" was his characterization in *Democracy and Education* (1916). The problem was that, as Lippmann had suggested, there was not yet a true public in existence to enjoy the benefits of this communication. Dewey was, as he sometimes said, an optimist about matters in the large, though gloomy about the state of the world at any particular moment—"a great optimist about things in general but a pessimist about everything in particular." One of the things he was optimistic about was the effect of mass communications technology. The radio, he thought, would lead to an increase in democracy, AND might create a truly communicative public.

The process was roughly supposed to be this. In a village, everyone learned of what was going on by a more-or-less instantaneous process of word-of-mouth transmission. He was fond of quoting from W. D. Hudson's portraits of village life, and of citing Hudson's account of the way in which an accident to a young man would become public knowledge within minutes of its happening. *The Public and its Problems* looked forward to a time when the intimacy and fellow-feeling of small-town America could be replicated on a national scale. The technologies of mass communication, especially the radio and the newspaper, seemed to him to be part of the necessary means to this goal. What he never imagined was what appears to have been the actual history of American mass communication from that time forward. In Britain, with a much more centralized system of government, and a large supply of high-minded public servants ready to give the public what it needed rather than what it wanted, the way was clear to a tradition of public service broadcasting that trod the narrow line between using radio as a means of state propaganda and allowing it simply to be used for mass entertainment. Sir John Reith's British Broadcasting Corporation was an inherently unstable solution to the puzzle of how to provide for the needs of highbrow, lowbrow, and middlebrow audiences, and how to have a coherent view of the purpose of public provision without having that view dictated by government. But it was a solution and it worked quite decently for sixty years.

There was never any attempt to create an American version of it. This meant that the United States has always provided a demonstration of the fact that technology is not an independent force, but rather an instrument of the local culture; contrary to Marshall McLuhan's great aphorism, the medium is not the message, but only a medium. American pluralism and commercialism got at the new media as soon as they were created, and their political impact has thus been what that pluralism and commercialism have dictated. For example, and to take a matter still on everyone's mind, it strikes everyone in the United States as extremely hard to control the amount of money spent on political campaigns. The reason is obvious enough; the medium through which voters are reached is television, and television advertising is extremely expensive. In most other countries, it is impossible to buy airtime for political advertising, and conversely, candidates for office either can advertise by piggybacking on their political party's advertising, or simply abstain. It is not the nature of television that dictates that this is how matters stand but the nature of the American commercial and political order. The consequence that is of the greatest interest for anyone who hankers after the Deweyan vision of a world where a plurality of different perspectives somehow unites a democratic public and provides it with a common opinion is that pluralism has turned into the enemy of democracy rather than its ally.

The situation is this. The proliferation of such things as cable television stations, and the uncontrolled posting of whatever one wants on the Internet, make for a certain sort of free expression. That is, anyone with access to the appropriate outlet can put into circulation whatever message he or she chooses. Anyone who has spent Sunday mornings trapped in a mediocre hotel will know what this means: that the television presents twenty or thirty different but depressingly similar variations on the theme of fundamentalist Christianity in full flight. Viewers are offered sexual fulfillment and financial security in return for a few dollars in contributions to the preacher on the screen—who usually has it in for gays, drug-dealers, socialists, secularists, or whomever, and needs our dollars to wage the holy war he has in mind. Now, if we were obsessed by free speech in the sense that every shade of religious opinion that could raise the money to broadcast itself to anyone who wished to see and hear it should be allowed to do so, we would find this a cheering spectacle. A thousand flowers would be blooming, and if some were a slightly odd color and shape, who is to complain of that? By the same token, we might think that the proliferation of wild ideas about the American political scene that one finds on the Internet is another manifestation of the varieties of human experience, and therefore something to be enjoyed and celebrated rather than to be deplored. This is not an absurd view, though it does presuppose something that is, alas, quite false—that these absurd religious and political ideas have almost no effect on the everyday life of the majority of Americans,

and that they can therefore be enjoyed as an aesthetic phenomenon. The problem is this: one of the ways in which Deweyan
democracy was supposed to work was by exposing the assorted,
discrepant, and plural views to the criticism of other views. Unlike
Mill and Tocqueville, Dewey was distinctively democratic in his
understanding of this process; where Mill minded that each of us
should come to have vivid, clear, and properly understood beliefs
of our own, Dewey was anxious that we should all contribute to a
common stock of intelligent belief in the hope that intelligent
action would flow from it. The appeal to "science" that has so
annoyed Dewey's critics over the years was not an appeal to the
findings of science, nor to the technological advances that science
made possible. It was an appeal to the practice of the scientific
community as a model for the practice of the democratic community. The trouble, in this light, with much recent communitarian talk is that it emphasizes the conditions that make for social
isolation and for the mutual reinforcement of nonsense. That is,
it concentrates on small communities and their moral solidarity;
Dewey wanted them to look outward to the wider American society, to be open to outside influence and to have a proper sense of
their connection to the larger democratic culture. He thought,
over-optimistically, that the radio and the newspaper would
achieve most of what was wanted; people could not help knowing
what was happening elsewhere because they would pick it up in
the air or off the page. It turns out that this is false. Indeed, it is
false twice over. For one thing that defeats the hope is that the
individual channels of communication are not aimed at those
who have no prior disposition to take notice of them. They are
targeted to an audience that is expected to receive what it is told
pretty uncritically. A magazine like *Guns and Ammo* is not published in the expectation that it will be read by Democrat-voting
assistant professors of English in small New England colleges. Its
design and intent is to reinforce the tastes and the spending
habits of men who enjoy shooting and who like to know more
about the weaponry they use. To the extent that it engages in

political discussion, it is entirely of the "we all know what motivates the other side" variety. As an invitation to dialogue it scores negative grades, which is hardly a reproach, since it never set out to make that sort of contribution in the first place. But what is true of *Guns and Ammo* is true to a great extent of absolutely everything on radio and television, and in the print media.

This may not be a very bad thing; or rather it certainly isn't a bad thing to the extent that what is being served is a special taste that in the nature of things we don't expect or want everyone to share. Nobody in their right mind would object that *Early Music* reinforces a taste for baroque and prebaroque music, does nothing to persuade jazz enthusiasts of the merits of Palestrina and Cavalli, and does not open up the question whether governments ought to steer clear of funding the arts. The anxiety we feel about the fact that local news programs may give a particular, narrow slant on what they feed to their audience is another matter. There we feel, or some of us feel, that the temptations to which local news stations succumb are politically dangerous. It is, for instance, a consequence of the combined effects of technology and finance that setting up a cable station is expensive, advertising is hotly competed for, and the need to grab an audience cheaply therefore is pressing. The easiest way to do this is to recycle old movies endlessly, but so far as news programs are concerned, the quick route is to fill slots with the raciest sort of local news, gathered by on-the-spot reporters using the cheapest possible equipment and presenting the material with a minimum of analysis. The result is, for instance, that the public believes that violent crime is on the increase when in fact it is dropping quite sharply. What they have seen is an increasing amount of violent crime on their television screens, and they can hardly be blamed for thinking that that is how the world is. But this is not a harmless misconception, since it affects the way they will vote, and the policies they will be willing to pay for. The impact of closed communal reinforcement of false views of the world is obviously much greater in the extreme cases, such as those of the various militias who convince themselves and each other that they are fighting against the forces of Satan; there, the estrangement from everyday reality is so acute that anything may happen, and as the Oklahoma bombing shows, sometimes what happens is very horrible.

I should recur again to the question of how far there is a technological imperative behind all this. I think there is not. There may be technological preconditions for the social structures to arise that make for this fragmentary and fissiparous kind of society—the rise of the automobile, the ease of bringing electricity to out-of-the-way places, the effectiveness of satellite television, and so endlessly on. It may be that the other effect of technology is what Richard Sennett and Robert Putnam among others have focused on: the thought that much, if not most, of the general information and political exhortation that we get will come to us as isolated individuals, passive consumers of what we take in. If that is so, then there is surely a risk that this unintegrated pluralism of ideas will get worse rather than better. Getting our information on a face-to-face basis tends to associate information with the people who are responsible for disseminating it. Having nothing of them present to us beyond their image on a television screen reduces the sense that ideas are something that people communicate to one another, and for which they can be asked to take responsibility. That leads to the curious, current situation in which people seem ready to believe everything and nothing simultaneously. They are both gullible and cynical; what they are not is properly skeptical, and properly ready to follow an argument where it leads.

As I say, the paradoxical thing about this position is not that it is surprising; it seems to me that most people share something like it. It is that it is a reversal of the old liberal anxiety about the impact of democracy on free speech. Nor is it a reversal in the sense of contradicting the anxieties of Mill and Tocqueville; the mindlessness that they feared so acutely is not dispelled merely by having a great variety of mindless opinions in circulation. Although they wanted the "antagonism of opinion" in order to keep democratic politics both free and lively, they also wanted that opinion to be focused on the task of getting coherent policies adopted by the government of the day, and having those policies accepted intelligently by the population at large. It was not antagonism for antagonism's sake that they had in mind. Of course, it is the Deweyan defense of pluralism that is most obviously the casualty of this new pluralism; his hopes for the harnessing of technology to an integrated pluralism have turned into a bad joke, and his expectation that plurality would lead to rational, egalitarian communication has been utterly disappointed.

Having said firmly that it is not technology alone that does the damage, but the way new techniques are used, and the commercial pressures that make for such use, I ought to end with one last observation about what "free speech" ought to mean in this context. I have deliberately avoided the great twentieth-century nightmare, by which I mean the vision expressed in Aldous Huxley's *Brave New World* (1946) of a world in which we were entirely unfree in our thinking because every deep-seated idea had been implanted before birth and immediately afterwards by a process of subliminal indoctrination. This, the theme of such postwar sociological works as *The Hidden Persuaders* (1957), strikes me as an overblown terror. Certainly, it is hard to believe that people voted for Ronald Reagan in any more rational a frame of mind than they usually choose their cars—but that is hardly to be complained of, since the question the electorate was posed was, "Do I like the way this guy talks?" and they found it easy to answer that they did. Their unswerving support for Congress's obstruction of almost everything he wanted to do suggests that the electorate mostly knew what it was doing in electing Reagan, and that we should not worry about their having been brainwashed. Nor have I spent any time worrying about the threats to free speech that are most obviously posed by technology—the technology of listening devices, surveillance cameras, the ability of governments to coordinate all the various databases that hold information about us, and all those things that would make it alarmingly easy for government to manipulate our fates and to make our lives impossible if it wished to. These strike me as genuine and obvious, and not sufficiently under control even in the United States. In Britain, the government appears to be bent on taking to itself the power to use them without let or hindrance, and the Labour Party is egging the government on. The thought that I think is worth having before us is a different one. It is that there are two sorts of freedom of speech or, if you like, two ways of connecting speech and freedom. One is encapsulated in the right to say pretty much whatever we like without fear of reprisals from the law. This is negative liberty in the old, familiar, liberal sense, and is well protected by the First Amendment. The other is less familiar, though it has recently been made much of in Michael Sandel's lugubrious tract on *Democracy's Discontent*. This is what he calls "republican" freedom, as opposed to liberal, laissez-faire freedom; it is the freedom that a community has when it is self-governing. A society can lose the capacity to govern itself for all sorts of reasons—general incivility, a loss of government capacity, a mass defection from the local loyalties; Sandel thinks that the United States is in a fair way to having lost that capacity because it has become morally chaotic. Nobody knows what life is about, and so nobody can get the public to rally behind anything in particular. This paper is in essence a footnote to such anxieties, the claim that traditional fears for the loss of our negative freedom of speech in an age of mass communication have turned out to be exaggerated, while our positive ability to speak to one another about matters of public moment seems for the moment to have decayed.

References

Bagehot, Walter, *Physics and Politics* (Gregg International, 1872).

Dewey, John, *Democracy and Education* (New York: Macmillan, 1916).

Dewey, John, *The Public and Its Problems* (New York: H. Holt and Company, 1927).

Huxley, Aldous, *Brave New World* (New York: Harper Brothers, 1946).

Lippmann, Walter, *Public Opinion* (New York: Harcourt, Brace and Company, 1927).

Maine, Henry, *Popular Government* (New York: H. Holt and Company, 1886).

Morris, William, *News from Nowhere* (Boston: Roberts, 1891).

Packard, Vance Oakley, *The Hidden Persuaders* (New York: D. McKay Co., 1957).

Putnam, Robert D., "Bowling Alone," *Journal of Democracy* 6:1 (January 1995): 65–78.

Sandel, Michael, *Democracy's Discontent* (Cambridge, Mass.: Belknap Press, 1996).

Williams, Raymond, *Culture and Society* (New York: Columbia University Press, 1958).

Constitutional Law and New Technology*

BY PAUL GEWIRTZ

How are new technologies assimilated into American constitutional law doctrines, particularly new communications technologies? I doubt that there is a single, generalizable process for this, true for all new technologies and all constitutional doctrines, but I do have a basic argument, and, in brief, it is this:

When our constitutional law is confronted with a new technology, two basic dynamics come into play. First, to some extent the courts simply seek to carry forward established constitutional principles in the new context. Constitutional rules governing the new technologies are developed by identifying the policies that established constitutional rules in their accustomed contexts, and deciding how best to further those policies as applied to the new technology. This is often difficult (the new technologies are both like and unlike the old, and the application of the established principles must reflect this); but it is the simplest understanding of what new technologies require of constitutional analysis.

But to some extent, a second dynamic often comes into play. The old principles are often not stable. They are being contested

---

* This is a light revision of remarks that I gave at the Conference on "Technology and the Rest of Culture" at the New School of Social Research on January 17, 1997. In March, after the Conference, the Supreme Court of the United States issued its second decision in *Turner Broadcasting System v. Federal Communications Commission*—a case that I had discussed in my remarks—and I have revised my discussion of that litigation accordingly. My January remarks also referred to the Communications Decency Act case involving the Internet, which the Supreme Court had just agreed to hear; the Supreme Court's decision was handed down on June 26, 1997, too late to be considered here.

on certain fronts. There is *normative* ferment in the air along with technological ferment. The new technology becomes the occasion to reconsider the old principles, sometimes covertly and even unconsciously. The new technology provokes new constitutional principles and suggests a different balance of policies, a new approach that has consequences for older technologies as well as the newer ones.

I. Electronic Wiretapping

Let me illustrate this process first with a comparatively simple example—how the Supreme Court dealt with the constitutional status of electronic wiretapping as that technology became widespread in the 1960s.

The main case is *Katz v. United States* (389 U.S. 347 [1967]). The constitutional problem was conceptualized in *Katz* as a Fourth Amendment "search and seizure" issue, not a First Amendment free speech issue, but the implications of electronic wiretapping for free expression are clear enough to justify this example here.

The specific constitutional question was this: Is an electronic wiretap on a telephone booth a "search and seizure" under the Fourth Amendment, triggering the standard Fourth Amendment requirement that a judicial warrant must first be secured?[1] The Court held that it was. The Fourth Amendment became part of the Constitution in 1791, when electronic wiretapping, of course, was an unknown technology—in fact when telephones were unknown. The framers of the Fourth Amendment, therefore, could not possibly be said to have intended the Fourth Amendment to limit electronic wiretaps of telephones. The Court's sole dissenter, Justice Hugo Black, concluded that the Fourth Amendment applies only to "tangible things" and physical intrusions, not conversations overheard by an electronic wiretap. "I simply cannot in good conscience give a meaning to words which they have never before been thought to have," Justice Black said, and "I will not distort the words of the Amendment in order to 'keep the

Constitution up to date' or 'to bring it into harmony with the times.'"

The Court majority analyzed the problem and interpreted the Fourth Amendment's general language differently. The Court reviewed its Fourth Amendment precedents and identified a general principle that underlies them: The Fourth Amendment "protects people, not places," and protects "the privacy upon which [a person] justifiably relied"; in the words of Justice Harlan's concurrence, it protects "reasonable expectations of privacy." Reasoning by analogy from prior cases where it had held the Fourth Amendment applicable, and recognizing "the vital role that the public telephone has come to play in private communication," the Court concluded that a person in a telephone booth justifiably relies on his privacy and is protected from electronic invasions as well as physical ones. To conclude otherwise "in the present day," the concurrence observed, would be "bad physics as well as bad law, for reasonable expectations of privacy may be defeated by electronic as well as physical invasion."[2] In short, the Fourth Amendment protects reasonable expectations of privacy, and if a new technology like wiretapping comes along that threatens that privacy in ways analogous to what is already protected by Fourth Amendment law, the Constitution is properly interpreted to cover those technologically newer dangers.

But, in fact, more was going on in the *Katz* case than simply applying settled constitutional principles to a new technology. *Katz* was decided during the heyday of the Warren Court, when the Supreme Court was rather generally redefining the balance between the rights of criminal defendants and the interests of society in strict law enforcement—redefining the balance in the direction of greater weight to the rights of the accused. *Katz* was part of this broader process of constitutional change. Pre-*Katz* Fourth Amendment law did not protect all "reasonable expectations of privacy." In fact, pre-*Katz* law had limited Fourth Amendment protections by making a series of distinctions that the dissent noted, generally not protecting conversations and protecting only against physical intrusions. Thus, the Fourth Amendment had long been held not to reach ordinary eavesdropping—a very old practice, and one obviously analogous to wiretapping. In extending Fourth Amendment protections to electronic wiretapping, the Supreme Court did more than extend the amendment to new technologies; it overturned old doctrines and distinctions that had limited the Fourth Amendment in other ways as well.

Thus, two developments here seem to me to go hand in hand. *Katz*'s extension of the Fourth Amendment to reach electronic wiretapping should be seen as part of a more general shift in our constitutional values during the Warren Court period, a more general expansion of the rights of criminal defendants. But *Katz* also reflected the Court's specific awareness of how newly vulnerable our privacy had become to electronic snooping, and how interpretation of the Constitution's protection of privacy had to extend to these new threats just to maintain equilibrium—that unless Fourth Amendment protection extended to new technology, the practical scope of protected privacy would end up being much narrower than it had long been. To some extent, in fact, the new technology was surely what prompted the broader reassessment of the distinction between physical and nonphysical intrusions and seizures of tangible versus nontangible items.

What we see reflected in *Katz*, in other words, is simultaneously an assessment of how a new technology might threaten traditional constitutional values, and also a reassessment and shift in those values. It is difficult to sort out precisely which strand of this dynamic is more responsible for the result in *Katz*. But my main point is that it is important to see both strands at work.

II. "New" Communications Media

This basic dynamic, I think, is also central to understanding how "new" communications media have been treated under the First Amendment, which requires that government "make no law . . . abridging the freedom of speech, or of the press." I first want to speak about the First Amendment in the context of government efforts to regulate media structure and access—that is, government efforts to enhance access to the media and thus enhance the diversity, balance, or robustness of public debate. Then I will look at government efforts to deal with particular *cultural* problems flowing from speech—namely, efforts to restrict "indecent" speech in new media.

A. *The Regulation of Media Structure and Access*

Let me start with some First Amendment theory. The dominant First Amendment model builds on the metaphor of a "marketplace of ideas," a metaphor tracing back to Justice Oliver Wendell Holmes' celebrated dissent in *Abrams v. United States.*[3] The model embodies faith in the speech market and distrust of government as a regulator of speech. The emphasis is on individual and media liberty. The expectation is that this expressive liberty will produce various social and systemic benefits, such as the rejection of bad ideas, encouraging a better public debate, and promoting a more deliberative democratic politics. But the focus of attention in the dominant First Amendment model is protecting individual liberty, and the government's duty is generally to keep hands off that liberty, not to regulate it to better achieve the hoped-for social benefits of the liberty. If there are imperfections in the speech market, well, there are thought to be far *greater* dangers in having government be the regulator of speech.

Broadcasting. This First Amendment model flowered in the era of print media, pamphleteers and soapbox orators, and First Amendment doctrine has developed in interesting ways when it has confronted some of the newer communications technologies, particularly broadcasting. The key Supreme Court case is *Red Lion Broadcasting Co. v. Federal Communications Commission* (395 U.S. 367 [1969]). At issue was the constitutionality of the Federal Communications Commission's fairness doctrine and personal attack rules, which required broadcasters to cover public issues in a balanced way and to give reply time to people they had personally attacked. The broadcaster in the case argued that these rules interfered with its First Amendment rights to broadcast what it chose, and to refuse, if it chose, to give a place to opponents' views.

The Court upheld the Federal Communications Commission in this case, concluding that "differences in the characteristics of [broadcasting] justify differences in the First Amendment standards applied to [it]." Most significantly for our purposes, the rationale the Court gives for its particular approach to broadcasting is based on technology: broadcast frequencies are limited and only a limited number of people can be allowed to broadcast if overlapping transmissions are to be avoided. So the government may license broadcasters and has distinctive power to impose regulations on those it licenses. Because of this spectrum scarcity, the Court said, "it is idle to posit an unbridgeable First Amendment right to broadcast comparable to the right of every individual to speak, write, or publish." The government may subordinate the First Amendment interests of the broadcasters to what the Court called the First Amendment interests of the "*public* to receive suitable access to social, political, esthetic, moral, and other ideas and experiences" (emphasis added).

This rationale for a special government power to regulate broadcasting—technological scarcity—seems questionable. All goods and resources are scarce. It is difficult to see why scarcity rooted in "technology" should be treated differently from scarcity for other reasons, such as "start-up costs," "market failures," or other economic factors. In addition, the technological scarcity argument has become increasingly difficult to maintain as stations proliferate on cable and other broadcasting media. Thus, another reading of *Red Lion* is possible. Perhaps it represents a more basic shift away from the dominant First Amendment model, a new approach to the First Amendment that is applicable not simply to broadcasting technology but to all free speech issues.

Under this alternative First Amendment approach, the government is allowed to intervene affirmatively to promote a better-functioning system of expression. The government would intervene not because of peculiarities of a particular *technology*, but because of recurring deficiencies in a largely unregulated "marketplace of ideas." And, in fact, *Red Lion* has been viewed this way by important First Amendment theorists such as Owen Fiss and Cass Sunstein (Fiss, 1996a, 1996b; Sunstein, 1993, 1995).

This alternative approach rests on a critique of the prevailing model of the First Amendment as applied to all media, not just to media defined by a particular technology. The most basic critique is that the market alone cannot deliver a sufficiently diverse and robust system of free expression. Market incentives of broadcasters and newspapers lead to inadequate information and debate. The high cost of speech limits the diversity of speech available to the public. Disparities of wealth mean that, in a largely unregulated speech market, wealthy speakers will drown out other speakers. Wealthy speakers, with viewpoints that are likely to reflect a distinctive band of interests, will have disproportionate influence in elections and broader public debate. All of this means that the public will not receive adequate information and hear a full diversity of robust opinions. In addition, the unequal status of certain groups, such as minorities and women, may make certain kinds of traditionally protected speech distinctively harmful to them, and therefore undeserving of constitutional protection—such as certain hate speech or sexual depictions of women.

Put more affirmatively, under an alternative model of the First Amendment, there are a variety of new justifications potentially available to justify government regulation concerning expression. These possible justifications for government regulation in the political sphere of speech parallel common justifications for government regulation in the economic sphere: there are frequent market failures in the marketplace of ideas just as in the marketplace of goods and services, and government regulation would be a corrective for these market failures. Or there are public interests that an unregulated speech market are unlikely to further adequately, including public interests rooted in the First Amendment itself, such as more diverse and open political debate. Or there are concerns about equality that an unqualified focus on individual liberty and the market may slight and that government intervention might address. Now, there are many, many problems with this alternative model that I am describing— most obviously, whether we can *trust* government to get involved affirmatively in improving the system of free expression, particularly when this sometimes means restricting some people's speech.[4] But there *is* this alternative way of understanding the First Amendment, and there are some suggestions of it in the *Red Lion* opinion.

So, then: Should *Red Lion* be viewed as simply a case resting on the technological differences of broadcasting, or instead read as a case representing elements of an emerging and normatively different model of the First Amendment? My own view is that both elements were at play in the case, and that at the time the case was decided it was possible that *Red Lion* would indeed be the progenitor of significantly reconfigured First Amendment principles. The year was 1969, and ferment was everywhere in political life and certainly on the Supreme Court. In the late 1960s, in one field after another, critiques of the market and consumer sovereignty were flourishing, and egalitarian perspectives also began to flourish and displace some classical individualist/liberty-based approaches.

But the broadest possibilities of *Red Lion* have not materialized since then. This seems quite consistent with other normative trends on the Court affecting much more than issues of speech and technology, a trend away from government regulation and toward the market, away from egalitarian insistence and toward libertarian acceptance, away from concepts of social rights and back toward viewing rights as individualistic. *Red Lion* is usually treated as a technology-driven case.

Most significantly, in *Miami Herald Publishing Co. v. Tornillo* (418 U.S. 241 [1974]), the Supreme Court refused to apply *Red Lion* to newspapers. *Tornillo* struck down a Florida statute that required newspapers to give a right to reply to political candidates they had criticized—an obvious parallel to the FCC's rule for broadcasting in *Red Lion*. There was no technological spectrum scarcity, of course. But the lower courts had upheld Florida's law in light of the actual scarcity of newspapers in most markets, and the public's interest in reading a wide variety of views. But none of this, the Supreme Court said, justified government regulation that would abridge a newspaper's right to publish what it wanted.

Cable Television. With cable television, however, the Court has been more ambivalent. From the perspective of a system of free expression, cable television is a revealing case study. It was originally hailed as a major technological breakthrough that would allow a much wider diversity of programming. And there is no doubt that cable has indeed produced some of this promised diversity and some outstanding programming. But the programming decisions are still overwhelmingly driven by market forces. This has meant vast quantities of recycled trashy sitcoms and low-budget trivial amusements, and wastelands of shock-talk and shopping-talk—Newton Minow's vista in wider panorama, vulgarized a good deal. It has meant news programs driven utterly by entertainment values that push aside serious analysis and instead emphasize personality-driven feature stories, lurid crime coverage, endless scandalmongering, and absurdly polarized and antagonistic pundit-debate. The public may want this programming, but that does not necessarily mean that the broader social purposes of the First Amendment are really being furthered. We have also seen that cable's vaunted diversity of programming has often deepened the segmentation of markets and the balkanization of the viewing public, with particular channels appealing more and more to a narrow band of the audience that watches these channels almost exclusively, gets exposed to few other perspectives, and has its preconceptions reinforced rather than questioned. And lastly, we have also learned that the diversity of cable programs is really all in the hands of the owner of a cable system, who controls the bottleneck through which programming must pass, and has effective monopoly power to decide what programming gets carried or does not get carried in any locality.

This last feature produced the congressional legislation that provoked the big showdown in the Supreme Court about the First Amendment's application to cable. The case is *Turner Broadcasting System, Inc. v. Federal Communications Commission* and involves the "must-carry" legislation of 1992, which says that most cable systems must carry local over-the-air broadcast stations whether or not they want to. Congress's basic concern was to assure the continued financial viability of the local over-the-air broadcasting stations that serve the 40 percent of the American public that does not subscribe to cable television. But cable system operators argued that the must-carry requirements violate their First Amendment rights by forcing them to carry programming they do not want to carry. In its first decision in the case in 1994, decided by a narrow five-to-four vote, the Court set forth its basic framework for analyzing the issue, but remanded to the lower courts for further fact-finding (*Turner I*, 512 U.S. 622 [1994]). In its second decision, issued this past March, also a five-to-four vote ruling, the Supreme Court held that the must-carry rules were constitutional (*Turner II*, 117 S.Ct. 1174 [1997]). Both prevailing opinions were written by Justice Anthony Kennedy.

Along the way to upholding the must-carry rules, the Court had to decide whether the technology of cable television is to be treated for First Amendment purposes like the print medium or like the broadcasters in the *Red Lion* case, and it came down rather ambivalently in the middle. In a key conclusion, the Court held that the First Amendment puts greater restrictions on government regulation of cable owners than on the over-the-air broadcasters in *Red Lion* because of technological differences in the media. The First Amendment approach in *Red Lion*, the Court said, was a product of spectrum scarcity in over-the-air broadcasting. But with cable there is no spectrum scarcity. The fact that cable operators had "bottleneck monopoly power" over access to the system was the not the same thing as spectrum scarcity, the Court said. On the other hand, the existence of this same "bottleneck monopoly power" of cable owners was part of the reason the Court gave for refusing to treat the government regulation here with the same "strict scrutiny" that it had used for the regulation of newspapers in the *Tornillo* case, where the newspapers did not have such a technological bottleneck.

This comparative assessment of technologies existed alongside extensive discussion of First Amendment policies, and here the Court revealed considerable normative ambivalence, if not normative ferment. The basic question for Justice Kennedy under the First Amendment was how to view legislation that restricted the speech of the cable operators in order to achieve the goals Congress had specified in the legislation: "(1) preserving the benefits of free, over-the-air local broadcast television, (2) promoting the widespread dissemination of information from a multiplicity of sources, and (3) promoting fair competition in the market for television programming" (512 U.S. at 662; 117 S.Ct. at 1186). If the Court had used the *Red Lion* approach or some variant of an alternative model of the First Amendment discussed above, it would have been rather simple to uphold the must-carry rules as a relatively modest effort by government to assure the broader viewer access to the television market and broader diversity of programming that Congress apparently had in mind in the first two of its objectives.

But Justice Kennedy clearly was reluctant to go down that path. "[T]he mere assertion of dysfunction or failure in a speech market, without more, is not sufficient to shield a speech regulation from the [strictest] First Amendment standards," Kennedy wrote. The dissenters in *Turner*, firmly adhering to the dominant First Amendment model, took the view that Congress's restriction of the cable operators' speech was "content-based" and therefore presumptively unconstitutional under current First Amendment doctrine. They cited considerable evidence in the legislation itself and the legislative record that what Congress meant by its first two objectives was promoting more diversity of viewpoints, more local news programming, more educational programming, and higher quality programming—all of which, the dissenters plausibly said, "make reference to content" (512 U.S. at 676), and amount to "a content-based preference" (117 S.Ct. at 1207). Far from defending any content-based efforts of this type, Justice Kennedy agreed that content-based regulation is presumptively unconstitutional but denied that any content preference was involved here. Instead he claimed that Congress was simply "recognizing that the services provided by broadcast television have some intrinsic value and, thus, are worth preserving" against the efforts of cable operators to "exploit their economic power" (512 U.S. at 648–49). Indeed, in both of his *Turner* opinions it is clear that he was trying as much as possible to justify the must-carry rules in light of Congress' more limited third objective, as a response to unfair anticompetitive behavior of the cable operators, rather than as a corrective for ordinary market forces that were jeopardizing a more robust speech market. But he does not abandon Congress's first two objectives. Justice Kennedy's opinions seem deeply divided between passages that appear to embrace a role for government in assuring "the widest possible dissemination of information from diverse and antagonistic sources" (117 S.Ct. at 1187), and others that reject government attention to the substance of what actually makes certain information sources "diverse and antagonistic." Normative uneasiness pervades the opinions.

Justice Stephen Breyer's concurring opinion in *Turner II* is worth particular mention. Justice Breyer joined the Court between the first and second *Turner* decisions, and his vote is the critical fifth vote needed to create a majority to uphold the law. His interesting opinion illustrates particularly well my basic argument about the interplay of technological and normative ferment: Breyer's acceptance of the must-carry regulation of cable television seems to reflect not only a response to the particular technologic features of cable television, but also a broader normative reconsideration of First Amendment doctrine.

Breyer, the newest Justice on the Court and a former Harvard Law School professor, appears to be rethinking First Amendment law. He has been noticeably reluctant in his first years to embrace the Court's standard doctrinal formulas in First Amendment cases. In a 1996 case involving "indecency" on cable (a case I discuss further below), Justice Breyer explicitly declined, at least in that context, to use any of the standard formulas to decide the case, instead engaging in a more direct balancing of interests, asking whether the statute "properly addresses an extremely important problem, without imposing, in light of the relevant interests, an unnecessarily great restriction on speech" (116 S.Ct. 2374, 2384–85 [1996]). He explained his doctrinal reticence in part by pointing to the technological flux in telecommunications. But the opinion as a whole suggests a sense of newfound freedom in not having to use doctrinal formulas to address a problem where "[t]he First Amendment interests involved are . . . complex, and involve a balance between those interests" (*Id.* at 2386).

Breyer's *Turner II* concurrence uses a similar methodology, but what is most revealing about Breyer's opinion is the substance of what he says about First Amendment principles and interests. He refuses to join Justice Kennedy's opinion "insofar as [it] relies on an anticompetitive rationale'," and relies instead on the rationale that the must-carry rules seek to provide "a rich mix of over-the-air programming," "to assure the over-the-air public 'access to a multiplicity of information sources,'" to promote "'the widest possible dissemination of information from diverse and antagonistic sources,'" and to prevent a decline in the "quality and quantity of programming choice." In embracing a positive role for government in "strik[ing] a reasonable balance between potentially speech-restricting and speech-enhancing" action, Breyer appears to be entering the territory of alternative First Amendment models (117 S.Ct. at 1203–1204).

I do not wish to overstate things: passages in Justice Kennedy's opinion have some of the flavor of passages in Justice Breyer's; Breyer significantly, although parenthetically, limits his argument by saying that government "speech-enhancing" intervention can be appropriate "at least when not 'content-based'" (although "at least" suggests perhaps also sometimes when content-based); and Justice Breyer focuses here "in particular" on the bottleneck feature of cable, and thus may not in fact be signaling a shift in First Amendment approach applicable more generally to deficiencies in the speech market. But what we clearly do see in Justice Breyer's opinion is an example of the dynamic I have emphasized: as new technologies toss up new constitutional issues to be addressed, considerable normative ferment about First Amendment principles often is at work at the same time.

Cyberspace. What about cyberspace? This term the Supreme Court is hearing a case involving so-called indecent speech on the Internet (about which I will say more below). But the Court has had no constitutional cases yet concerning the basic structure of access in cyberspace, and, in fact, it is quite unclear what issues might come up about this.

One reason is that the basic "architecture" of this new world— the configuration of technologies, market structures, social organization, and government regulations—is not yet set (Berman and Weitzner, 1995). And given this, it is hard to know for sure what cyberspace will produce for our political life or for our system of free expression, or what "structure and access" problems will actually end up developing. But given the quite exuberant predictions that seem everywhere these days—and are embellished in Arno Penzias's article in this issue—I want to make a few observations.

Cyberspace does offer many appealing things for a system of free expression that seeks to be "uninhibited, robust, and wide-open"—Justice William Brennan's memorable phrase in *New York Times v. Sullivan* (376 U.S. 254, 270 [1964]). There will be easy, low-cost, speedy access to huge quantities of information, and easy opportunities for exchange of views among large numbers of people. Both the scarcity problem with over-the-air broadcasting and the bottleneck problem with cable television are nonexistent here, since no one owns the Internet. There will be considerable parity among speakers. Quite simply, more people will be speaking more, with more back and forth exchange. In our political life, there may be more debate, more new candidates who are now priced out of campaigns, more input by citizens to their government, more communication by government officials to the public—and therefore a more immediately responsive government. As indicated above, current First Amendment doctrine basically ignores the effects of the high cost of speech for the functioning of the system of expression, and this has produced problems and concerns that have pushed some to favor a substantial retooling of the doctrine. But, as Eugene Volokh has observed in a brilliant article, since the new cyberspace technologies seem to be decreasing the cost of speech and fostering a wide-open market of speech and debate, *current* First Amendment doctrine, unmodified, may "work better than it ever has before" (Volokh, 1995, p. 1807).

But I remain at least cautious about cyberspace's effect on our system of free expression and political life. A new ease of access to information seems undeniable, and pretty clearly beneficial. I am less sure about our gains from the interactive dimensions of cyberspace. To give just one example, with respect to political life, there may be more written input and access to the offices of government officials, and more responses ostensibly from those officials; but we all know that the officials themselves typically won't read most of it or actually write the responses. And the volume of junk mail will hugely rise.

The egalitarian promise of cyberspace may also be exaggerated. More well-to-do people may be communicating, but the poor may be left even further behind, both because the start-up costs are high (you need a computer, software, and a on-line service) and the skills needed are quite high.

Cyberspace also seems to encourage an extreme form of the balkanization and segmentation of markets that I mentioned in connection with cable television. The positive side of scarcity and limited broadcast options in the other media is the possibility of a common culture in which people are exposed to diverse viewpoints. With scarcity, if you want to watch television you have to watch channels that necessarily try to reach a broad audience, and therefore usually have somewhat diverse programming; you have to read a newspaper that is read by many other people and has lots of different kinds of news and opinion in it. One of the major possibilities of cyberspace, however, is that it will allow much more customization of what you read (Volokh's article is particularly visionary here). People may tailor their access to get just the stories they want. Indeed, as people have less and less need to leave their computer screen to go to an office, a school, a mall, even a meeting, the nature of interpersonal and group interactions in a communal space may be transformed. True, those who want diverse news and freewheeling chat rooms can get that too, and of course people will still get out and around. But there's a real possibility that cyberspace may erode or at least radically change the idea of a common culture and may greatly reduce people's exposure to viewpoints that challenge their own.

To the extent that cyberspace promises more "direct democracy"—more direct input by the people unfiltered by their representatives—that too could be a serious problem. The founders of our Constitution deliberately set up a representative democracy rather than a direct democracy so that informed and deliberative representatives would make the day-to-day political judgments, with the people an ultimate, rather than daily, check. Of course, developments in mass communications have always provoked concern in some quarters about runaway populism. But cyberspace opens up some truly unique dangers. Cyberspace makes

Ross Perot's electronic town hall technologically possible, a terrible prospect in my judgment, since it would allow poorly informed and unreflective popular opinion to decide every "legislative" issue. American political culture has already softened up the ground for the darker possibilities of these technological developments. Public opinion polls are relentlessly used to supplement or displace expert assessments of complicated technical matters, though those being polled often lack enough knowledge of the relevant facts. There has also been what I call an Oprahfication of much of American life—we're all now like Oprah's studio audience, where everyone's judgment is deemed valid and equally weighty regardless of one's knowledge, and where the implicit credo is "I feel, therefore I may judge." We put ourselves on a par with, or above, "experts," and there is broad distrust of all of our representative institutions and elected officials—sometimes well-founded, of course. But until now, the simplistic populism that many have muttered didn't have to be delivered because it *couldn't* be delivered. Cyberspace technology, which allows instant polling and rapid input and responsiveness, makes wide-scale direct democracy possible for the first time, and could powerfully transform our political life—very possibly for the worse.

Having noted these potential dangers, however, it is doubtful that government regulation will provide an answer.[5] Basic characteristics of the Internet make it likely that government regulation of Internet speech (unlike broadcasting speech) will be evaluated under the stricter doctrinal standards that exist in our current First Amendment law, reflecting the dominant First Amendment model. Nor is it apparent where, if anywhere, the currently competing First Amendment models and theories would point to government regulation of cyberspace *structure* beyond antitrust-type interventions and government subsidies for the basic hardware and software in schools and elsewhere, which the dominant First Amendment model itself comfortably allows. Cyberspace is developing at the moment in such an "open access" way that those who have pushed for regulation to address access problems in other media seem to have little reason at the moment to regulate cyberspace structure, and many reasons not to. The wiser course, however, is probably to say that it is simply too early to tell either how Internet structure will evolve or how constitutional understandings stimulated by that structure will ultimately shake out.

B. The Regulation of "Indecent" Speech

The other large area of First Amendment law that I want to discuss concerns government's efforts to regulate speech that is "indecent" or "patently offensive" but not necessarily "obscene." The regulatory impulse here has little to do with concerns about the basic *structure of access* to the media and how that structure might affect democratic political life. Rather, the concern here is typically with the *cultural effects* of certain kinds of expression— here, speech related to sex that is not pornographic and that many people find valuable or otherwise enjoyable but that many people consider offensive, or harmful to the general cultural climate, or harmful to children.

The Supreme Court has generally held that books and magazines that meet quite specific definitions for obscenity and pornography can be prohibited, but not those that are merely "indecent" or "patently offensive." Might different and more restrictive standards apply to the technologies of broadcasting, cable television, and the Internet?

The key case, once again, concerns broadcasting: *Federal Communications Commission v. Pacifica Foundation* (438 U.S. 726 [1978]). As you may know, the Supreme Court upheld a Federal Communications Commission decision punishing Pacifica for a daytime radio broadcast of a George Carlin monologue called "Filthy Words," in which various indecent language was repeatedly used. We have already seen in the *Red Lion* decision that the spectrum scarcity of over-the-air broadcasting is invoked to justify special government efforts to assure *greater* access for people or viewpoints that otherwise would not be heard on the scarce frequencies. Spectrum scarcity would be harder to invoke to justify *prohibiting* certain speech content on the airwaves.

But *Pacifica* points to other particular characteristics of broadcasting to justify the restrictions. The first is that the broadcast media have established a "uniquely pervasive presence in the lives of all Americans." "Pervasiveness" here seems to suggest that greater government regulation over broadcasting is justified because broadcasting has come to have such a dominant place that it defines our cultural life. This probably does explain much of the concern that people have about the content of broadcasting. Popular entertainment in visual and audio media saturates our culture, sweeping the country along in trends and attitudes that few of us can resist and creating a cultural climate that affects much of what we do as well as our sense of happiness and well-being. Too often, defenders of free speech ignore this culture-transforming dimension of speech and the profound "harms" many people perceive that it causes. But acknowledging such "harms" does not mean embracing censorship as the solution. Most relevant, if these concerns were enough to justify content regulation of broadcasting, then the government would have vast power to restrict broadcaster freedom, a much more vast power than court decisions have ever suggested. Restrictions on expression have been allowed to protect against certain immediate harms that expression can cause, but not to prevent unwanted cultural change from pervasive forms of speech as such.

Another particular feature of broadcasting that the Court is that it "intrude[s] on the privacy of the home without prior warning as to program content." The concern, apparently, is that in the course of listening to a program, or drifting through a room where the television or radio is on, or simply changing stations, a listener is vulnerable to hearing things that he or she may not want to hear. But much the same situation of coming upon material we do not want to experience surely arises with all media, not just broadcasting.

The other characteristic of broadcasting that the Court discusses is that broadcasting is "uniquely accessible to children." This is the characteristic that seems most important to the Court. With books, parents can hide away sexual material that they do not want their children to see, and booksellers can be barred from selling certain material to children while remaining free to sell the material to adults. In other words, children can be protected "without restricting the expression *at its source*" (emphasis added). But broadcasting is different, the Court said, because radio and television broadcasts are readily accessible to children, and their intrusiveness adds to the difficulty of shielding children. With broadcasting, therefore, the government has greater power to restrict expression "at its source" because it is harder to limit children's access at the distribution point (a car or living room).[6] The scope of that power is not unbounded, but the FCC had limited its restrictions to certain times of the day, and the Court says that the indecent speech here might be broadcast in the "late evening" when few children would be in the audience.

Now all of this suggests that *Pacifica* should in fact be seen as a rule for broadcasting, reflecting the Court's best attempt to adjust First Amendment principles specifically to this medium. But once again, I think it essential to see that the Court's decision also seems to reflect more general normative ferment about First Amendment principles. The case was decided in the 1970s, at a time of deep concerns about the continuing effects of the 1960s counterculture and sexual revolution, and deep concerns about a decline of American cultural standards and civilities. (The new Chief Justice after Earl Warren's retirement, Warren Burger, made some of these concerns the subject of judicial and extrajudicial writings.) It is reasonable to think that the *Pacifica* case reflects a new willingness by some members of the Court to reconsider whether First Amendment law was protecting too much expression that was degrading social and cultural life. A few years before *Pacifica*, in 1973, a new Court majority undertook a full-scale reexamination of obscenity law under the First Amendment, and significantly narrowed the First Amendment protections for all media. And *Pacifica* itself introduced other related modifications of First Amendment law that applied much more broadly than to broadcasting, and were probably responsive to the cultural concerns I have just mentioned.

Most importantly, a plurality of the Court embraced a theory of "lower value" speech that applied to "offensive" speech of the sort involved here, and applied to such speech in *every* medium, not just broadcasting. Such speech is not entirely outside the protection of the First Amendment, like obscenity; but it is given much watered down protection because of its "lower value." For example, the government is allowed to sharply limit the "location" of speech that it might not be able to prohibit altogether. In *Pacifica*, the limited location is certain restricted time-slots, a feature relevant to broadcasting. But in cases since *Pacifica*, with "lower value" speech in other media such as movie theaters showing "adult" films that are not obscene, the Court has also allowed similarly restrictive regulation of location, allowing local governments to require the segregation of adult movie houses into very circumscribed places (see *Renton v. Playtime Theaters, Inc.*, 475 U.S. 41 [1986]). In short, for all of its seeming technology-specific character, *Pacifica* also represents broader normative changes in First Amendment doctrine—changes that the technological setting may have *crystallized*, but that are not specific to it.

But, interestingly, some recent technological developments may be helping to rescramble First Amendment law a bit once again. In the Supreme Court case that has already considered some of these developments involved cable television, *Denver Area Educational Telecommunications Consortium, Inc. v. Federal Communications Commission* (116 S.Ct. 2374 [1996]), Justice Breyer's opinion declined to use the familiar doctrinal formulas of First Amendment law. *Denver Area* addresses the constitutionality of some recent congressional regulations of "indecent" and "patently offensive" programming on cable television. One provision of the statute, which the Court upheld, permitted cable operators to completely bar indecent or patently offensive programming from leased access channels on their cable systems. Another provision, which the Court struck down, required cable operators who allowed such programming to segregate it on a single channel, block it from viewer access, and unblock it within thirty days of a subscriber's written request.

A variety of factors persuaded a very splintered Court to uphold the first provision. One important factor was that, even though the provision restricted communication opportunities for certain programmers and viewers, it actually promoted the First Amendment rights of the cable operators themselves, since the legislation gave them the power to decide whether or not to carry indecent or patently offensive programming. But another key factor in the case was the Court's conclusion that the cable broadcasts here are as "accessible to children" as the over-the-air broadcasts in *Pacifica*, and therefore the cable technology posed the same difficulties in protecting children from indecent programming.[7]

But the Court struck down the segregate-and-block provisions of the act. This section not only created restrictions for viewers and programmers, but also directly interfered with the First Amendment interests of cable operators (whose First Amendment interests had been protected in the other section of the act). The Court concluded that these segregate-and-block requirements were unconstitutional because they were not tailored narrowly enough to the objective of protecting children. And why weren't they tailored narrowly enough? Because of technological developments.

Less restrictive means, the Court said, were now available to protect children from offensive material on cable—namely V-chips, lockboxes, and blocking without requiring written requests. It was the availability of these "less restrictive" methods for protecting children—methods made possible by technological advances—that led the Court to strike down the more restrictive segregate-and-block provisions on First Amendment grounds.

Thus, I think we see here the centrality of some interrelated technological factors in applying First Amendment rules about indecent speech to newer communications media.[8] First, different communications technologies have different basic availabilities for children. We all know how easy it is for children to access television, how much time they spend doing it, and how difficult it has been for parents to monitor what children are watching. The easier children's access is, the easier it will be for government to justify regulation.

Second, in some settings new "filtering" technologies will permit parents or programmers to easily and effectively block programs from reaching children without blocking adults (Balkin, 1996). The less costly and burdensome these blocking technologies are, the harder it will be for government to justify flat-out prohibitions. It is yet to be seen whether the V-chip will be easy for parents to use and whether their more high-tech children will be able to outsmart them (and, to mention a low-tech question, whether the rating system on which it will rely can be applied effectively to the huge number and variety of cable broadcast programs). But the V-chip does seem to be a promising technological route out of some of the problems being considered here, a route that would allow even wider availability of certain programming to adults while allowing parent-imposed (rather than government-imposed) restrictions on children's access to this programming. So, contrary to what some civil libertarians have said, the V-chip may end up enhancing our civil liberties by making it constitutionally impossible for government to altogether ban or even "safe harbor" certain programming; and it may also end up enhancing the ability of parents to control the programming their children see.

We have already seen this in the cable television case. And we are likely to see parallel arguments in the important case involving the Internet that is before the Supreme Court this term, the first case in which the Court is considering the Internet (*Reno v. American Civil Liberties Union* (929 F.Supp. 824 [E.D.Pa. 1996],

probable jurisdiction noted, 117 S.Ct. 554 [December 6, 1996]).
The Supreme Court is considering the constitutionality of the
Communications Decency Act (CDA), federal legislation that
makes it criminal to transmit or display to children on the Internet material that is "indecent" or "patently offensive" (even
though not obscene). The main constitutional attack on the law
is that as a practical matter it suppresses the transmission and
display of lots of material to adults. Again, a pivotal part of the
First Amendment argument is likely to be whether technologies
exist or are likely to exist soon that will make it possible for parents or program providers to block access by children to this
material without blocking adults.

The judges on the three-judge federal District Court that struck
down the law found that there was no technologically and economically feasible way for most providers to avoid liability by
screening for a user's age; that tagging their transmissions so that
users could themselves filter out unwanted material was currently
only a hypothetical possibility and would not avoid liability
anyway; and that user-based software was not yet available that
could do the filtering, although "reasonably effective" software
"will soon be widely available" (929 F.Supp. at 838–41, 845–48,
855–57). The government argued that if the CDA was upheld, it
"will likely unleash the 'creative genius of the Internet community
to find a myriad of possible [technological] solutions'" (929
F.Supp. at 857).

The crucial and interesting question is what should be the constitutional significance of these facts, assuming they are true (and
remain true when the Supreme Court decides the case). If there
were a feasible technological way for providers to screen for a
user's age, then arguably the act's defenses adequately protect
providers and the act would be constitutional. If there were a feasible technological way for users themselves to screen for
unwanted material, then the act would appear not to be the least
restrictive means for protecting children and therefore would be
unconstitutional (as in *Denver Area*). If there is currently *no* feasi-

ble technological way to filter out the offending program for children, then one can imagine arguments either way: the sweeping prohibitions of the CDA might therefore be seen as the "least restrictive means" presently available to protect children and thus be constitutional. Alternatively, if the absence of filtering technology means that providers have no way to avoid liability with respect to children except by ending transmissions to adults too, then that could be viewed as an unacceptably heavy restriction on speech. (This was the District Court's conclusion.) A further possibility, given some factual uncertainties about the ever-evolving filtering technologies, is that the case will turn on who has the burden of proof: if, as seems right, the Government has the burden of proving that no less restrictive methods than the CDA are available to protect children adequately, then the Supreme Court might perhaps conclude simply that the government has not met its burden given the apparent promise of user-based filters.

I leave you with a closing restatement of my theme, which may actually be a somewhat modified statement: there are undoubtedly evolving "cultural" concerns that help shape legal outcomes—here, cultural concerns about proliferating possibilities for sexually explicit speech. But technological change also affects legal analysis significantly, providing both arguments for and arguments against First Amendment protections in new contexts.

Notes

[1] The Fourth Amendment to the United States Constitution provides that: "The right of the people to be secure in their persons, houses, papers, and effects, against unreasonable searches and seizures, shall not be violated, and no Warrants shall issue, but upon probable cause, supported by Oath or affirmation, and particularly describing the place to be searched, and the persons or things to be seized."

[2] Justice Louis Brandeis had made a prescient argument along these lines in his famous dissent in *Olmstead v. United States,* 277 U.S. 438, 471–85 (1928).

3 250 U.S. 616, 630 (1919) ("the basic test of truth is the power of the thought to get itself accepted in the competition of the market").

4 A significant short critique is Krattenmaker and Powe, Jr., 1995.

5 Regulation, of course, may focus on issues other than structural ones—issues such as pornography and indecency, privacy, encryption, criminality—which raise different constitutional questions.

6 The *Pacifica* case itself involved a mid-day radio broadcast of the "Filthy Words" monologue, which brought a complaint from someone who heard the broadcast while driving with his young child.

7 Recall that in the context of *access* questions, the Court has treated government regulation of cable more strictly than regulation of broadcasting because of spectrum scarcity in the broadcasting (512 U.S. 622, [1994] supra). But cable is no different from broadcasting in its basic accessibility to children, which was the government's focus here; the spectrum scarcity difference was irrelevant.

8 This paragraph and the next are much indebted to my colleague Jack Balkin's marvelous article, "Media Filters, The V-Chip, and the Foundations of Broadcast Regulation" (1996).

References

Balkin, Jack, "Media Filters, The V-Chip, and the Foundations of Broadcast Regulation," 45 Duke L. J. 1131 (1996).

Berman, Jerry, and Weitzner, Daniel J., "Abundance and User Control: Renewing the Democratic Heart of the First Amendment in the Age of Interactive Media," 104 Yale L.J. 1619 (1995).

Fiss, Owen M., *The Irony of Free Speech* (Cambridge: Harvard University Press, 1996a).

Fiss, Owen M., *Liberalism Divided* (Boulder: Westview Press, 1996b).

Krattenmaker, Thomas G., and Powe, L. A., Jr., "Converging First Amendment Principles for Converging Communications Media," 104 Yale L.J. 1719 (1995).

Sunstein, Cass R., *Democracy and the Problem of Free Speech* (New York: Free Press, 1993).

Sunstein, Cass R., "The First Amendment in Cyberspace," 104 Yale L. J. 1757, 1759–1765 (1995).

Volokh, Eugene, "Cheap Speech and What it Will Do," 104 Yale L.J. 1805 (1995).

Abrams v. United States, 250 U.S. 616 (1919)

Denver Area Educational Telecommunications Consortium, Inc. v. Federal Communications Commission, 116 S.Ct. 2374 (1996).

Federal Communications Commission v. Pacifica Foundation, 438 U.S. 726 (1978).

Katz v. United States, 389 U.S. 347 (1967).

Miami Herald Publishing Co. v. Tornillo, 418 U.S. 241 (1974).

New York Times Co. v. Sullivan, 376 U.S. 254 (1964).

Olmstead v. United States, 277 U.S. 438, 471-85 (1928).

Red Lion Broadcasting Co. v. Federal Communications Commission, 395 U.S. 367 (1969).

Reno v. American Civil Liberties Union, 929 F.Supp. 824 (E.D.Pa. 1996), probable jurisdiction noted, 117 S.Ct. 554 (December 6, 1996).

Renton v. Playtime Theaters, Inc., 475 U.S. 41 (1986).

Turner Broadcasting System, Inc. v. Federal Communications Commission (Turner I), 512 U.S. 622 (1994).

Turner Broadcasting System, Inc. v. Federal Communications Commission (Turner II), 117 S.Ct. 1174 (1997).

PART 6

Imagination

What are the effects of changing technologies on modes of representation and ways of imagining? How have philosophers, artists, and writers in the past and in the present responded to these changes?

Introduction

BY ROSALIND WILLIAMS

THIS entire journal issue could appropriately be dedicated to the memory of Lewis Mumford (1895–1990), who set forth its fundamental theme in *Technics and Civilization:*

> No matter how completely technics relies upon the objective procedures of the sciences, it does not form an independent system, like the universe: it exists as an element in human culture and it promises well or ill as the social groups that exploit it promise well or ill (1934, p. 6).

Technology as part of human culture, not its antithesis; technology as an expression of imagination, not its enemy: these principles still need to be articulated and defended. Indeed, they are more important than ever before. The most important technological development in the last half-century—the development of the computer—offers particularly striking confirmation of the thesis that technology is part of culture, not an objective, independent procedure. Computational technology emerged from the quantitative calculations of ballistics trajectories, and was initially applied to the quantitative calculations of corporate bureaucracies; by the 1980s, however, computers rapidly migrated into communications and entertainment, and by the 1990s, the binary rationality of computer technology was associated as much with e-mail, web-surfing, and games as with functional number-crunching.

If this journal issue, then, is one long footnote to *Technics and Civilization,* this section offers a more particular and more ominous set of insights into the relationship between technology and culture. All these papers make a similar series of connections: technology as an expression of the imagination; the imagination as the irrational; the irrational as violent, aggressive, and destructive; and therefore technology as violent, aggressive, and destructive. These connections are especially emphasized by Herbert, who shows how, in the modern visual arts, technological images are routinely associated with military ones. In science fiction, surrealism, dada, photo-collage, and many other arenas, artists were trying (as Herbert quoted Raoul Haussman during the conference) "to snatch out of the chaos of wartime and revolution an optical and thoughtful new reflected image."

In short, these papers repudiate the conventional wisdom that technological images and themes express primarily logic, utility, functionality, and rationality. Instead, they emphasize what George Steiner has termed "the imagination of disaster," and its connection with technology. This connection is only too obvious, considering this century's repeated history of technological development in service of mass destruction (as Mumford himself came to realize by the time he wrote *The Myth of the Machine* [1967]).

In his paper, however, George Kateb refines this argument in two steps. First, he proposes that the rational and the irrational are not opposites, that the motivation behind the most apparently rational technological projects can be deeply irrational. Second, he proposes that this motivation involves a peculiarly Western hatred of the natural world, a hatred that manifests itself in a deep anger at the world as it is given and an equally deep desire to distance humanity from nonhuman creation.

Kateb's arguments suggest some interesting ways of analyzing the history of technological development. For example, if brought to bear on the postwar history of the computer, they make the story more complex than the evolution from computational rationality to communicational and playful irrationality described above. His analysis emphasizes, first, the irrational drives that were there from the beginning, in the form of the military and economic passions that were served by computing machines. Second, his argument highlights the extent to which modern computers are used to escape from earthly constraints to the apparent freedom of the placeless digital universe.

One of the high virtues of this section is the way that it calls attention to the vocabulary we use to talk about these developments. As Hollander so rightly points out, the very concept of *technology* is a feat of the imagination. He notes that "only literature . . . can represent technology itself." He notes that modern historiography engendered the very concept, and that before the industrial age the Latin-rooted term *art* carried rather different associations than the Greek-based *techne*.

Elsewhere in this journal issue, Leo Marx pushes this analysis even further, arguing that the very concept of *technology* as an abstraction is dangerously alienated from the world, as it implies an impersonal agency and therefore distances human beings from the effects of what we are actually doing. In that case, it is tempting to call in the imagination as a corrective to the alienating and evasive lure of technology. But just as Marx argues that there is no singular, static entity of technology, maybe there is no such singular category of the imagination. If the Western imagination is death-centered and world-alienated, is it not possible to imagine an imagination that is eros-centered and world-loving? The texts and works of art described in these papers are for the most part well-known products of the Western canon from classical to modern times. Much has been omitted, most notably music and film. Furthermore, we could consider scientific as well as artistic imagination; collective as well as individual; popular as well as high art; and so on.

An essential role of artistic and literary imagination is to reground us in earthly creation, to call us back from sterile abstractions to the thickness and richness of existence, both human and nonhuman. But if world alienation is such a fundamental part of Western consciousness, then will not our imagination as well as our technology be affected by that outlook? In that case, our imagination will only reinforce, not counteract, the aggression and world alienation of the technological project. The fundamental question of course is whether imagination can only express preexisting wants and needs, or whether it is powerful enough to have the capacity to rise above itself, to imagine new wants and needs, to imagine another relationship with non-human nature.

References

Mumford, Lewis, *Technics and Civilization* (New York: Harcourt, Brace, and World, 1934).

Mumford, Lewis, *The Myth of the Machine: Technics and Human Development* (New York: Harcourt, Brace, Jovanovich, 1967).

Technology and Philosophy

BY GEORGE KATEB

L ET us begin with a general definition of the word *technology*. In its current meaning, it names the means or methods used to help people move from place to place, communicate, produce, construct, create, fabricate, but also destroy; to observe, to calculate, and to think. (This list is obviously not exhaustive.) Technology is thus made up of all kinds of equipment—tools, machines, and devices—that assist the work of human muscles, senses, and brains, and thus the realization of human purposes and ends. Our question is: Is *modern* technology a subject of philosophical interest? Philosophers can make any subject interesting, and not just to themselves. At its best, their work has a tendency to arrest our habit of taking things for granted, of seeing phenomena as normal or as a matter of course. Philosophy, from Socrates on, is often a challenge to common sense. Philosophy is amazed at the way in which common sense is unamazed; and then in another sense of amazement, philosophy is disposed to be amazed at the phenomena themselves. Philosophers have sometimes characterized our steady condition as sleep; their hope is to awaken those who hear or read them. Thoreau said that "Moral reform is the effort to throw off sleep" (Thoreau, 1937, p. 81); so too is the reform of one's perception. To be awake is to experience amazement repeatedly; it is to marvel. I submit that modern technology is worthy of amazement, and that some philosophers have tried to awaken us to amazement.

I refer, however, not to the prowess, the marvels of technology: we are fully awake to them, whether as active users or passive consumers. Instead, I have in mind the marvelous fact, the amazing fact, of modern technology as a human relation to nature and

283

human beings. There is nothing ordinary, universal, or inevitable about that relation. Naturally, I do not mean to belittle anyone who looks with amazement at any given example of modern technology or at a set of connected examples. The feats of modern technology are staggering; the capacity for ever more feats in the future seems intact. Yet, this prowess, this capacity, grows out of some passions or drives or motives that repay attention; they are culturally and historically special. Philosophers, especially German and American ones, have tried to uncover them, and have done so, marveling not only at the feats and triumphs of modern technology but also at the mentality and condition of spirit that launched modern technology, and have helped to keep it in its tremendous course of continuous inventiveness and continuous revolutionizing of technique.

The whole meaning of modern technology cannot of course be contained in any single philosophical system or tradition. Nor is philosophy needed on one level of explanation. There are, after all, countless consumerist pleasures, active powers, and special interests served by the development of modern technology; there are also countless accidents and strokes of fortune that went into its making. No particular story about modern technology could be adequate, no interpretation sufficient. But some thinkers have a powerful contribution to make to our understanding of modern technology; they add depth.

But, first, what is the *unamazed* common-sense attitude toward modern technology? I hope that I am not too reductionist when I attribute to common sense the view that modern technology is just extraordinarily successful problem-solving, a terrific display of resourceful ingenuity, a splendid and constant show of adaptability to circumstances. Of course, problem-solving, ingenuity, and adaptability are traits essential to the story of modern technology. But what originally called forth these traits? What has helped to call them forth repeatedly, once modern technology is launched? Why have they appeared more in the West than any place else? Why have they appeared in such profusion only in the period that begins with the Italian Renaissance and the German Reformation—the period that many scholars call modernity? Both modern science and modern technology begin in that period—not from scratch, of course, but in any case substantially. Whatever drove the beginnings of science and whatever drives it still, the career of modern technology is, however, not wholly subsumable under that of modern science. As Heidegger says, modern technology is not simply applied science, although it is that, too, of course (Heidegger, 1977, p. 116). The will to work technologically on nature and human beings is conceptually independent of the will to know for its own sake, which is characteristic of science.

I suggest beginning with Karl Marx's views. He is surely one of the first to think philosophically about modern technology, especially machine technology housed in factories. The question that I think should be put to his work is whether he philosophically challenges our common sense or, instead, expends his genius in keeping us asleep. Who could doubt that his work aroused many people from the 1880s on, and then aroused a large part of the world after his death? His vision seemed to have awakened people to their place in the world, to their true identity and interest. He powerfully insisted, what is more, on the distinctiveness of modernity. Yet I believe that this great philosopher actually helped to strengthen common sense in its mental sleep, rather than awakening it to amazement. What I mean is that the tendency of the Marxist system is to see modern technology as an intense concentration of human practicality, not as an achievement stemming from passions, drives, and motives that enlist the desire for success in solving problems in a much larger and rather mysterious project.

Marxism does deserve tribute for insisting on the way that technological capitalism, especially in the period from the middle of the eighteenth century onward, had transformed Europe and affected the rest of the world. In "The Communist Manifesto" (1848), Marx and Engels say of the European bourgeoisie:

It has been the first to show what man's activity can bring about. It has accomplished wonders far surpassing Egyptian pyramids, Roman aqueducts, and Gothic cathedrals; it has conducted expeditions that put in the shade all former Exoduses of nations and crusades (Marx and Engels, 1978[1848], p. 476).

Marxism thus points to the marvels of modern productive technology, and does so comparatively early and certainly with compelling rhetoric. Just by that effort, carried out in much of their writing, and culminating in *Capital* (1867), Marx, and Engels, too, perform one of the duties of philosophy in the face of common sense.

But something large is missing. I mean that Marxism offers only a little help and poses an enormous obstacle in the attempt to give an account of why it was the West that launched modern humanity on the career of technology. Marxism's insistence on the role played by individual and class economic self-interest explains something about the persistent use made of technology by modern capitalism, but next to nothing about the mentality of its emergence. When we turn to the most ambitious foray into the philosophy of history and historical change made by Marx (with help from Engels), we are given a much too narrow base for our speculation. I refer to a passage in *The German Ideology,* a work written in 1845–46 (and then withheld from publication). Marx says there that unlike other German philosophers who habitually explain historical change by reference to philosophical innovations, changes in ideas and conceptions, a true realist must begin with certain physical or biological premises. These premises must organize the writing of history; they must provide the basic cause of the epochal transformations in the human condition that have taken place from time to time, and with greater speed and perhaps greater self-consciousness in the past few centuries. Nothing psychological or spiritual or philosophical is needed to make sense of history.

What are these premises that provide the key to understanding historical change? Marx gives three (Marx and Engels, 1978

[1848], pp. 155–57). The first is that in order to live and go on living, human beings must engage in some kind of labor. The second is that "the satisfaction of the first need . . . leads to new needs" (p. 156). That is, wants expand, even if their satisfaction is not required to keep bare life going; and these expanded wants are felt as imperiously as needs for food, dress, and shelter. The artificial becomes as necessary as the natural. The third premise is that male and female must engage in sexual intercourse if the race is to go on. Thus, production and reproduction constitute the foundation of human existence. What gives this truism its intended force is the Marxist contention that much of the time people carry on their various activities unaware that the real, but largely hidden point or aim of all their activities, is to keep life going. Noneconomic and nonsexual activities have the same ultimate point or aim as economic and sexual ones, or are ultimately at the service of economic and sexual ones. Humanity is driven by the mission of life, even though it often believes that it is inspired by other and more flattering purposes. All its ideas, conceptions, and philosophical innovations have only one source, if mostly hidden: the imperatives of the mission to keep life going. No aspect of human life has freedom from these imperatives, even though people characteristically delude themselves into thinking that their creations, their culture, their various activities are not only severally autonomous, but are also the unconstrained and unpredictable source of their particular and variable definitions of how to respond to nature, how to respond to the mission's undeniable imperatives. Marx is saying that even inequality and class domination have simply been the best instruments of these imperatives.

Marxism wants us to marvel at the wonders of capitalist productivity, but not at their psychological or spiritual or philosophical causes. These wonders have occurred, Marxism suggests, because humanity is always trying to increase productivity, always serving economic rationality, even when it thinks it is doing something else. The dialectical change from one economic system to another is, for all its violence and drama, a steady story of change: growth in productivity. Thanks to an accumulation of technical knowledge, which is inevitably sought, human productive capacities have gradually increased, despite setbacks, through time; but thanks also to some fortunate circumstances, these capacities have grown with a relentless acceleration from the time of the European discovery of the New World. Marx is not exactly saying that modern technology was *fated* to happen. He is saying, however, that modern technology, propelled by the self-interested profit-seeking of the entrepreneurial class, has served the imperatives of the mission of keeping life going, and until recently has done so better, more fully and efficiently, than any conceivable alternative. In short, the relation to nature and human beings that modern technology manifests is nothing to wonder at. Human beings have always had the same relation: they must produce in order to keep life going, where life requires that strict necessities of food, dress, and shelter, but also the acquired necessities that developed productivity creates, be satisfied. Marxism elaborates both the old adage that necessity is the mother of invention, and to some extent, also elaborates the reverse, which is that, as Thorstein Veblen puts it, invention is the mother of necessity.

Other philosophers challenge the contention that humanity has always had the same relation to nature and human beings. They see in modern technology something in addition to improved productivity, and hence something in addition to the old story in which humanity does battle with natural scarcity in the most effective way possible (even if that way, to be effective, had to be ruthless and unfair). They also see something besides the reign of what neo-Marxists have called "instrumental rationality," the pursuit of means to means, with ends forgotten. There is a zeal that lets nothing stand in the way of ever-greater efficiency in the production of more and more goods, and all this for the sake of ever-greater profits, and in total disregard of the costs to workers or to nature, while all higher purposes recede, dwarfed

by the technological process. Despite the suggestiveness of the
critique of instrumental rationality, however, there is another
kind of amazement at this process. A number of philosophers
express it. What do they see?

I will refer only to three thinkers who, explicitly or not, take
issue with Marxism and with the common-sense attitude to modern technology that Marxist philosophy has helped to fortify, or
even to create. These three are Max Weber, Martin Heidegger,
and Hannah Arendt—all German, or German in origin.

Now, I do not say that trying to awaken us to amazement should
be the only office of philosophy. There is intellectual advantage
when some philosophers, and Marx is one example, side with
common sense and embolden it. Other philosophers go in the
other direction and produce the ideas that help to change people's relation to nature and social reality, and thus recreate common sense altogether. Marx gives an example, also, of this
interpretative power. In this war among the offices of philosophy,
a given philosopher can move from one side to another, from
time to time. Let us say that the archetypes are Socrates, the gad-
fly; the Sophists, who defend and abet common sense; and Plato,
the lawgiver (even if a devious one). For the purposes of this
paper, Weber, Heidegger, and Arendt are gadflies: they try to sting
us into wonder about modern technology; while Heidegger also
emerges as a kind of lawgiver or prophet, but more negative than
positive.

I will be very selective in discussing the philosophical challenge
to common sense and to Marxist and neo-Marxist philosophy on
the matter of modern technology. What holds Weber, Heidegger,
and Arendt together is their inclination to regard modern technology as stemming from the passions and drives and motives of
excess or extremism, not from resourceful practicality. In the
background is a great page from the Third Essay of *On the Genealogy of Morals* (1887), where Nietzsche says, "measured even by the
standards of the ancient Greeks, our entire modern way of life,
insofar as it is not weakness but power and consciousness of

power, has the appearance of sheer *hubris* and godlessness." We violate nature and "we cheerfully vivisect our souls" (Nietzsche, 1969[1887], sec. 9, p. 113). Not only are the feats of technological prowess, in their profusion, a cause for wonder, the passions and so on underlying them are the amazing heart of the story and must be attended to, if the feats themselves are to be properly appreciated and properly marveled at.

I find in Weber, Heidegger, and Arendt the following themes, all of which are first stated or suggested by Weber in *The Protestant Ethic and the Spirit of Capitalism* (1958[1904–5]). All the themes go against both philosophical Marxism and against common sense. First, it is a fact worthy of meditation that, in its origin, modern technology is not a human-species-wide phenomenon, but a distinctively Western one. Although the West was and is indebted to the scientific and technical achievements of other regions and cultures, it is responsible for modern technology. In the three thinkers, and others as well, the Western distinctiveness is rendered ambivalently: celebration and profound disquiet are mixed together in the analysis of both the feats of prowess and the underlying passions that brought those feats into existence. I would add that such ambivalence is one of the most prominent features in a lot of the writing about technology.

A second theme is that, just as modern technology was not a universal phenomenon, so there was nothing inevitable about it in the West. All three philosophers have a keen feeling for accident and contingency, for roads not taken, for opportunities either not accepted or forced into being. A third theme is that the passions, drives, and motives that helped to promote technology are, to a significant extent, instigated or inspired by ideas, religious or philosophical, that manage to be absorbed into the self-conception that many individuals in a cross-national cultural setting hold of themselves. Modern technology is not only applied science; even more profoundly, it is philosophy or theology enacted. Philosophers and theologians are the originators. Other people go along or are carried along, because of the original reasons, or because of their own varied reasons, half-reasons, and nonreasons.

The long and short of it is that only in the West, but contingently so, modern technology emerged, and did so because of the birth and spread of hard, abstract ideas that are not explainable in the Marxist manner as inevitable reflexive responses to material circumstances. And it has emerged and continues to flourish as a special and particular project; what is more, a project of excess and extremism. The modern West, best seen as the creator of modern technology, is not only distinctive, it is anomalous in comparison to the rest of the world; and the anomaly is stupendous and perhaps monstrous. The marvel lies not only in the results of the project but, to begin with, in the fact that the project of modern technology was ever undertaken and then sustained. Humanity in the West has had a certain kind of relation to nature and human beings not to be found elsewhere, or found in a much diluted form.

Now, Weber's aim, to start with him, is not to account for modern technology, but to explain the rise of capitalism. It turns out, however that for Weber capitalism is only one, though a major, expression of Western distinctiveness. Common to all the features by which the West has distinguished itself from the time of the Greeks onward is rationalism, rational method, a continuous displacement of superstition and wishful thinking in regard to nature and human beings. Every aspect of Western culture—its architecture and art, its political systems, its science, its scholarship, even its religions—has been characterized by rationalism. Now, rationalism is a rather elastic term in Weber, no doubt. But his accumulation of examples adds up to one of the best brief overviews of Western distinctiveness ever attempted.

Capitalism is the example that occupies most of Weber's attention in *The Protestant Ethic and the Spirit of Capitalism*. The role of capitalism in developing technology, and also the role of technology in developing capitalism, may not figure with much explicitness in Weber's analysis. He is primarily interested in the preindustrial origins of the spirit of capitalism, the period of the sixteenth to the eighteenth centuries. What matters, however, is the *affinity* between capitalism and technology, not their causal connection. They partake of a similar attitude to reality. Both are manifestations of what Weber conceptualizes as rationalism, rational method in any given sphere of life. Hence what Weber says about the rise of capitalism, we may plausibly apply to modern technology.

But what is the connection between rationalism and that excess and extremism to which I have referred? Rationalism would seem to be the very opposite of excess or extremism, and if culturally anomalous, then certainly not monstrous. Yet this is where Weber's analysis, and those of Heidegger and Arendt as well, take on a heightened interest. The contention is not simply that too much rationalism—rational method after a rather distant point— turns against itself and becomes inhuman, though that thought appears in Weber. Rather, the contention is that the passions or drives or motives that push rationalism everywhere and to an apparently limitless extent are themselves not rational, but irrational. Reason in the form of rational method is at the service not so much of everyday basic interests as of a philosophical or religious outlook on life or reality, or the world, or nature and human beings. Weber's thought is more vividly expressed earlier by Herman Melville's Captain Ahab in *Moby-Dick:* "all my means are sane, my motive and my object mad."[1]

What I wish to propose is that in distinct contrast with, or in explicit and sharp opposition to Marxism, a number of German thinkers as well as a number of earlier American writers have thought they glimpsed rational method in the service of something not always distinguishable from madness. Method in madness, for the sake of madness, but also the rational method itself, by its unrelenting quality, often veering into madness, and thereby partaking of the nature of the end. Ahab's formulation does not cover the whole subject; it may not even cover his own experience. All that the narrator of *Moby-Dick* credits Ahab with is "some glimpse" of the truth about himself. A little later (Ch. 46, "Surmises"), the narrator speaks of Ahab's "subtle insanity" and "strange imaginative impiousness." The excellence of Melville's analysis is that he takes pains to distance Ahab from his crew. Ahab is aware that love of the chase is an "evanescent" feeling among the crew. "The permanent constitutional condition of the manufactured man, thought Ahab, is sordidness." By sordidness is meant only plain self-interest, which can be enlisted, however, in a leader's enterprise, an enterprise that is so self-absorbed that the usual motive of self-interest or even love of the chase does not originate or drive it. Of course, Ahab is not an inventor, nor a symbol of modern technology; but his injured and outraged stance towards life strongly resembles the composite of drives and passions and motives, of obsessions and mental leaps, attributed to modern technology by some philosophers.

Indeed, rational method, when sufficiently uninhibited, looks as if it were the end itself, so intensely gratifying is its use, while the ostensible end (whether rational or irrational) appears to be a pretext. Edmund Burke supplies the political analogy when he describes political revolution: "criminal means once tolerated are soon preferred. . . . Justifying perfidy and murder for public benefit, public benefit would soon become the pretext, and perfidy and murder the end" (Burke, 1970, pp. 176–77). Yet what he says about revolutionaries can be said about many other political actors, for whom political office is an attractive invitation to use criminal means in conscienceless good conscience. So, to a certain extent, with modern technology. The very distinction between extreme means and extreme ends in modern technology is volatile, so regular is the transformation of each into the other.

From a certain perspective, at least, modern technology often looks like madness, or like nonhuman giantism, to use Heidegger's concept (Heidegger, 1977, p. 135). There must be madness when modern technology shows itself at least as dramatically in destruction as in productivity. One is tempted to say that modern technological weapons of war are not, to those who invent or design them, command or use them, qualitatively different from the means of economic production for civilian uses. The psychology is the same in both cases. How could we fail to have our sense of wonder quickened when we observe that the technological spirit passes so easily and quickly from peace to war, as if there were no humanly real difference between them? To find in that move mere practicality, ordinary problem-solving, is to be lost in a dream, just as those who set the "problems" to be solved are in the grip of hubristic ambitions. The more important point, however, is that something like madness—call it irrationality—sets modern technology in motion and helps to keep it in motion.

In Weber's analysis there would have been no modern capitalism, but also no modern science or (we may add) modern technology, without something irrational. This is not to say that for Weber Western rationalism has been implicated in irrationality from the start. But he does wonder at the thought that the West has applied rational methods to areas of life left comparatively untouched by such methods elsewhere. These methods appear irrational to those outside the West, and they also appear irrational to some Western philosophers.

Weber locates the source of modern rational irrationality in the Protestant idea of work as a calling, the practice of work as worldly asceticism. The tremendous energies of modernity were unleashed not by worldly self-interest, by prudence, calculation, or the concerted effort to deal with needs that are either primary or added to primary ones, or even by the high aim of relieving the human estate, but rather by systematic self-denial. If the world were run by a more conventional self-interest, or even by a decent commitment to human betterment, it would be an easier and less productive place, with less wealth but less poverty, less misery. Greed exists everywhere, Weber says, but the rational and methodical pursuit of profit is the essence of capitalism, and distinctive to it. And apart from the causal connection between capitalism and modern technology, each phenomenon, considered separately, is driven by the application of rational method to limitless purposes, whether limitless profit or limitless technical achievement. Both purposes signify the presence of the inexhaustible, the insatiable, the unappeasable.

But why the limitless? In regard to capitalism, as we have said, Weber makes much of the Protestant notion of one's work as one's calling. Thus, Protestantism, especially Calvinism, shifts the notion of vocation or calling from religious practice to the everyday work done by all persons in the world. Why the shift? Weber says

> the religious valuation of restless, continuous, systematic work in a worldly calling, as the highest means to asceticism, and at the same time the surest and most evident proof of rebirth and genuine faith, must have been the most powerful conceivable lever for the expansion of that attitude toward life which we have here called the spirit of capitalism (Weber, 1958[1904–5], p. 172).

By a curious alchemy, which Weber does not satisfactorily delineate, but persuasively emphasizes, anxiety over salvation or damnation led persons to find both distraction from anxiety and palliation of it in unremitting work. Success in work gave some possible clue to one's fate in the afterlife. Worldly blessings were an arbitrary sign of God's favor, not a loving expression of it. Success in work, at the same time, went to those who applied the most rational method to their work, while they disciplined themselves to labor much harder than most people, including their fellow Christians. They were driven most powerfully not by economic interest but by a religious conception that was not rationalist and not even rational, but an affront even to many Christians. In a word, religious anxiety lies at the root of capitalism understood as a special case of the method of rationalism. The tremendous tower of economic achievement rests on a base that is not worldly, but religious. Only a passion not worldly, not itself rational, apparently, could inspire such stupendous worldly results, but could achieve them only by a wholehearted use of rational methods.

We can say that if religious anxiety helps to account for the emergence of capitalism in the West, something like it may help to account for the emergence of modern technology in the West; and, furthermore, something like it may help to account for the distinctive rationalism of the West from the Greeks onwards. Weber's analysis of the sources of a compulsion to use rational method cannot be confined to Protestantism and capitalism, as he indicates by his prefatory consideration of Western distinctiveness, even if he withholds the theme of the irrational basis of rational method until he discusses capitalism.

I suggest that Protestant anxiety, now present with us in the form of a secular work ethic, is just one manifestation of a general anger at the world that seems to exist in the West much more than elsewhere, and from the Greeks onwards. I do not say that anger at the world is the only passion or drive or motive that underlies modern technology—the technology of production and destruction. But it is one of the few main elements. (I will briefly mention some others at the end.) And when I speak of anger at the world I include anger at oneself and the tendency to project or imagine divinities that are given to anger at us. There is a recurrent disposition in the West to feel radical dissatisfaction with nature or reality or the world; to wish it otherwise; to find fault with the given to the point of undying regret that things are as they are; and, along with all this, to look with perpetual discontent at oneself and strive to improve oneself by making oneself an object of perpetual surveillance and solicitude.

Of course I highlight only one part of the moral temper of the West. It is the part that Weber and later German philosophers highlight. I believe that they have caught something noteworthy that should figure in the effort to explain Western distinctiveness and with it, modern technology. Early in the second chapter of *Walden* (1854), "Where I Lived, and What I Lived For," Thoreau says

> Why should we live with such hurry and waste of life? We are determined to be starved before we are hungry. Men say that a

stitch in time saves nine, and so they take a thousand stitches today to save nine tomorrow (p. 83).

In the sentence "We are determined to be starved before we are hungry," I find a formulation that exceeds in significance the particular economic point that Thoreau is making. He is mindful of the "desperation," as he elsewhere puts it, that characterizes modern life. We make ourselves sick so that we "may lay up something against a sick day" (p. 5). (To be sure, he is prepared to admit that he has sometimes been a "half-starved hound" ranging the woods, "not that I was hungry then" [p. 189].) He laments the general situation. Whether we lament it or not, the insight remains: we cannot seem to make sense of the Western anomaly unless we perceive modern technology as flowing not from need or lack but from fullness of energy. And as Thoreau suggests in a spirit not remote from Weber and post-Weberian philosophers, fullness of energy breeds anxiety, anger, dissatisfaction, and self-dissatisfaction, and they reciprocally increase energy. Modern technology comes not from hunger: we are determined to be starved before we even feel hunger. Satisfied need does not appease the psyche. We cannot believe our own fullness, or, we refuse to accept fullness as fullness. In that fullness we are determined to act not merely as if we are hungry but as if we are starved. We are starved no matter how much we eat. We are also eager for the opportunity of emergency, or necessity. We pretend to a bare need when what we suffer from is an excess of determination—a determination to use up energy in anger, anxiety, dissatisfaction, and self-dissatisfaction. And what we find is that energy accumulates with its expenditure. Success provokes dissatisfaction. What I am submitting to doubt is the contention that the modern technological project has been instigated by a memory or anticipation of starvation—a determination never (or never again) to be starved. Rather, indefinite hoarding ("standing reserve" in Heidegger's phrase) is a preparation for some future breakthrough, some magical moment of use, some deed that leaves the world unrecognizable (Heidegger, 1977, pp. 3–35, 36–49).

All this is not solely Protestant; and if modern technology is *modern* technology, there is nevertheless a perfect consonance between modern technology and salient features of the Western spirit from the Greeks onward. It is impossible to say whether religion and philosophy originate these features of the Western spirit. But at the least, religion and philosophy in the West have codified and sanctified and thereby emboldened anger, anxiety, and so on. The ultimate cause may be buried too deep in obscure and accidental psychic processes, past and present, for us to unearth.

Heidegger and Arendt amplify the story told by Weber, and also told, well before him, by Thoreau and Melville and others in the United States. In such essays as "The Age of the World Picture" (1938), "The Question Concerning Technology" (1949–50), and "Science and Reflection" (1954) (all three in Heidegger, 1977), Heidegger finds the origin of modern technology in Western metaphysics. He emphasizes the outlook of thinkers as lawgivers, and sees in them the true bearers of the passions and drives and motives that are the foundation of modern technology, rather than assigning the inspiration to common Western humanity as a whole. When philosophers are not content to awaken wonder at the world, but instead strive to remake the world, they sometimes succeed. Their greatest success is modern technology. Modern technology is, to repeat our phrase, a certain relation to nature or reality or the world; it is therefore not merely the inevitable application of that immense scientific knowledge that grows once humanity is rid of communal or religious superstition and repression. Certainly that is part of the story, but for Heidegger it is not the principal part. Rather, modern technology is the materialization of Western metaphysics, which is the parent of both modern technology and modern science. Indeed, the technological aim drives the development of modern science. Western metaphysics is just one interpretation of reality, just as its offspring—modern science and modern technology—are particular relations to reality, to what is given.

Heidegger means to show that Western metaphysics—and metaphysics includes theology—is a continuously if sometimes covertly reiterated Platonism. By his method of exegesis, Heidegger tries to persuade us that Platonic metaphysics converts the world into a picture for the mind's eye, and by doing that, prepares Western humanity to lose sight of the mere fact of existence, the unsummoned thereness of reality, of the given. Metaphysics inveterately reduces the world. The purpose of the reduction is to make the world intelligible and hence manageable, fit to be worked on, and made ready to have practical order imposed on it. The world, as given, is disliked; it is disliked in large part just because it is given; the dislike engenders anger, and from anger comes rebellion. Western humanity is and has always been at war with given reality, to a much greater degree than the rest of humanity, and in a remarkably distinct manner. Technology is the most spectacular campaign in the great war waged by Western humanity against nature or reality as given. To repeat: the deepest cause of that war is not scarcity, not the failure of nature to make better provision for a necessitous humanity, but, instead, a Western willfulness, a will to power, to mastery, an overflow of energy that wants to shake the world to pieces and make it over. The craving is either to put the human stamp on reality or at least to rescue nature from the absence of any honestly detectable stamp, any detectable natural purpose or intention. As Nietzsche says: humanity, in its *asceticism*, "wants to become master not over something in life but over life itself, over its most profound, powerful, and basic conditions" (Nietzsche, 1969, sec. 11, pp. 117–18). Western humanity cannot let things be on their own terms or coax gently from them their own best potentiality; it is so far unable to practice what Heidegger calls *Gelassenheit*. Western metaphysics is the sponsor of anger and hence of repeated violence towards nature.

Hannah Arendt, in *The Human Condition* (1958) and other writings, continues the philosophical effort to restore a sense of wonder at the very fact of modern technology, at the relation to reality that modern technology stands for. She speaks of world alienation in the modern age, a process that she dates from the discoveries of Copernicus and Galileo and their impact on the thinking of philosophers. In her writings, I find two sequences. In one, world alienation begins in those discoveries that discredit the senses and force the skeptical mind back on itself to find its warrant for certainty there rather than in the testimony of inadequate senses concerning the deceptive world. There are other events. The Renaissance and Reformation combined with the European invasion of the New World to produce a profound disorientation, a loss of worldly assurance, a loss of the feeling that it is possible to be at home in the world. The world and its possibilities are made too vast. From such alienation, resentment grows, and with resentment the project of making the world over, and then persisting in undoing and remaking it forever. Resentment is the key concept for Arendt; it is much broader than Nietzsche's *ressentiment,* but has some of the latter's qualities.

Again, it is not resentment at nature's scarcity, but at its sheer otherness that drives the project of modern technology, which is the project of mastery and domination. There is in fact an alternative and superior theoretical sequence in Arendt's writings. Arendt leaves the impression that Western humanity perhaps never wanted the world to be a home, and that if it were more congenial to human expectations, Western humanity would have been even more alienated, more resentful at the given, more disposed to inflict a technological simulacrum on it. In this account, resentment is prior, certainly among thinkers. Western thinkers have always resolved to be alienated from the world. Resentment spurs the developments that culminate in pervasive modern world alienation, which then further intensifies and spreads and, as it were, popularizes resentment. She shares with Heidegger a disposition to see the ancient Greeks as already living for the sake of imposition on otherness: their metaphysics and their science seem animated by the same, or roughly the same, mentality as modern science and technology.

This is not to say, however, that Arendt loves nature and the natural. She has her own snobbery. She is repelled by all those human beings whose labor (especially agricultural) makes them, in her eyes, indistinguishable from natural creatures who are totally absorbed into nature's "metabolic" processes. She loves existence as such, she loves the world, but she does not seem to love the natural. (Compare this to Thoreau, who loves the natural, the primitive, but is disgusted by the sensual and the sexual.) The climax of Arendt's story is in her essay, "The Conquest of Space and the Stature of Man" (1968[1963]), in which she claims that the upshot of space exploration will be the flight from the earth and the colonization of space (pp. 265–80). World alienation could thus lead to a radical earth alienation. From space, human beings will look down on their earthly habitation from a distance both physical and metaphysical and thus consummate their restless resentment at earthly and worldly limitation. "Modern rationalism is unreal," she concludes, "and modern realism is irrational" (Arendt, 1958, p. 300).

Common to Heidegger and Arendt is the wish to promote a feeling of astonishment that there is a world that is just there, given without necessarily there being a giver; and beyond astonishment at givenness, grateful adherence, or attachment. Gratitude would show itself in, among other ways, an acceptance of limits on the will to rebel against nature. To try to subdue it to the satisfaction of human needs is inevitable; the process must go on, if human life, and not only life but civilization, are to go on. It is the Western rebellion at all limits, the anger or resentment that givenness provokes, the imperious rage to remake the world so that humanity becomes the only maker, even in destruction—all this is not compatible with gratitude for existence and perpetually substitutes wonder at human achievement for wonder at the fact of existence itself.

"We murder to dissect," says Wordsworth in his poem "The Tables Turned" (1798). We pretend to understand the living thing when we study it dead, after we have killed it. The word *murder* indicates Wordsworth's judgment that only the murderous would dissect, and that dissection is but a continuation of the act of murder. Hatred or dislike of the creature or natural thing is behind the dissection—a burning desire to master it by knowing it when it is no longer itself.

But the question arises as to where a genuine principle of limitation on technological endeavor would come from. It is scarcely conceivable that Western humanity—and by now most of humanity, because of their pleasures and interests and their own passions and desires and motives—would halt the technological project. Even if, by some change of heart, Western humanity could adopt an altered relation to reality and human beings, how could it be enforced and allowed to yield its effects? The technological project can be stopped only by some global catastrophe that it had helped to cause or was powerless to avoid. Heidegger's teasing invocation of the idea that a saving remedy grows with the worst danger is useless. In any case, no one would want the technological project halted, if the only way was a global catastrophe. Perhaps even the survivors would not want to block its reemergence.

As for our generation and the indefinite future, many of us are prepared to say that there are many things we wish that modern science did not know or is likely to find out and many things we wish that modern technology did not know how to do. When referring in 1955 to the new sciences of life, Heidegger says

> We do not stop to consider that an attack with technological means is being prepared upon the life and nature of man compared with which the explosion of the hydrogen bomb means little. For precisely if the hydrogen bombs do *not* explode and human life on earth is preserved, an uncanny change in the world moves upon us (1966, p. 52).

The implication is that it is less bad for the human status or stature and for the human relation to reality that there be nuclear destruction than that (what we today call) genetic engineering should go from success to success. To such lengths can a mind push itself when it marvels first at the passions, drives, and motives that are implicated in modern technology, and then marvels at the feats of technological prowess. The sense of wonder is entangled with a feeling of horror. We are past even the sublime, as conceptualized under the influence of Milton's imagination of Satan and Hell.

It is plain that so much of the spirit of the West is invested in modern technology. We have referred to anger, alienation, resentment. But that cannot be the whole story. Other considerations we can mention include the following: a taste for virtuosity, skill for its own sake, an enlarged fascination with technique in itself, and, along with these, an aesthetic craving to make matter or nature beautiful or more beautiful; and then, too, sheer exhilaration, a questing, adventurous spirit that is reckless, heedless of danger, finding in obstacles opportunities for self-overcoming, for daring, for the very sort of daring that Heidegger praises so eloquently when in 1935 he discusses the Greek world in *An Introduction to Metaphysics* (1961, esp. pp. 123–39). All these considerations move away from anger, anxiety, resentment, and so on. The truth of the matter, I think, is that the project of modern technology, just like that of modern science, must attract a turbulence of response. The very passions and drives and motives that look almost villainous or hypermasculine simultaneously look like marks of the highest human aspiration, or, at the least, are not to be cut loose from the highest human aspiration.

Notes

[1] Melville, 1851, Ch. 41, "Moby Dick." Leo Marx instructively discusses this passage in *The Machine in the Garden* (1964), pp. 298–302.

References

Arendt, Hannah, *The Human Condition* (Chicago: University of Chicago Press, 1958).

Arendt, Hannah, "The Conquest of Space and the Stature of Man," in *Between Past and Future*, 2d ed. (New York: Viking, 1968).

Burke, Edmund, *Reflections on the Revolution in France* (New York: Penguin, 1970).

Heidegger, Martin, *Discourse on Thinking,* John M. Anderson and E. Hans Freund, trans. (New York: Harper, 1966).

Heidegger, Martin, *An Introduction to Metaphysics,* Ralph Manheim, trans. (New York: Doubleday Anchor, 1961).

Heidegger, Martin "The Age of the World Picture," in *The Question Concerning Technology and Other Essays,* William Lovitt, trans. (New York: Harper, 1977).

Marx, Karl, *Capital,* Ben Fowkes, trans. (New York: Vintage, 1977 [1867]).

Marx, Karl, and Engels, Friedrich, "Manifesto of the Communist Party," *The Marx-Engels Reader,* 2d ed., Robert C. Tucker, ed. (New York: Norton, 1978).

Marx, Leo, *The Machine in the Garden* (New York: Oxford University Press, 1964).

Nietzsche, Friedrich, *On the Genealogy of Morals,* Walter Kaufmann, trans. (New York: Vintage, 1969).

Thoreau, Henry David, *Walden,* in *Walden and Other Writings* (New York: Modern Library, 1937).

Weber, Max, *The Protestant Ethic and the Spirit of Capitalism,* Talcott Parsons, trans. (New York: Scribner's, 1958).

Literature and Technology: Nature's "Lawful Offspring in Man's Art"

BY JOHN HOLLANDER

MY rubric may be thought to cover at least three different questions, each of them vast and complex. The first concerns the technologies contributing to and shaping the formation of the activities and institutions of literature throughout history. The second comprises the matter of technology as represented in or by literature. I shall be more concerned with second of these than the first. A third question, as to whether literature is itself a technology, will be considered shortly.

A trivially technological response to being asked to discuss literature and technology might be to inquire—in the language by which many of us cope with databases—what sort of Boolean operator the "and" in this case is? As a rubric it could cover: *Literature "about" Technology; Literature "as" Technology; Literary Technologies* (whatever they might be); *The Literature of Technology* (also ambiguous). The first question, as to what contrivances enable literature to exist, lurks behind these: all of them seem initially to have been generated by the first technologies with which speech was represented and thus partially, preserved. It might be noted that this embraced both the hardware of making enduring marks on surfaces, and the software of encrypting language in various forms of writing system. A technological history would observe such sequences as pictogram-syllabary-alphabet as well as tablet-papyrus, vellum-roll-codex (and back to virtual roll on the computer screen), for example.

The question of writing systems themselves might extend to the crucial matter of printing and the current matter of the microchip (which may or may not prove to be as, or more, crucial), and so forth. I mention this only because it seems inevitable to start with the implication that literature is itself a technology. The notorious charge leveled by Socrates in the *Phaedrus* against the technology of writing, and how inventing it supplanted and ruined the earlier, better, and somehow more natural operations of memory (the nineteenth century might think of this as being more "organic"), suggests that the very invention of writing was a new technology whose product would be what we call literature. But I shall leave moot here what is to me the most interesting question of oral preliterature and whether building discursive structures out of phonological materials prominent in particular languages—what Roman Jakobson called the "organized violence" committed by poetry on the ordinary probabilities of occurrence of phonological features—can or should itself be designated technological. Is, then, the music—the linguistic, syntactic, and rhetorical rhythms, even occasionally the choreography—of primitive song the fruit of a kind of primal software? I would think so, but shall consider oral literature only for a moment in passing here.

The development of writing itself, then, particularly as it has been investigated in the past two decades by the late Eric Havelock (1986) and Jesper Svenbro (1993), might be conveniently stipulated as the initial enabling technology. And yet, the identification of the literary with what is written aside, the matter of "oral literature" raises another question here. It can be assigned, as Plato implicitly does, to some notion of a pretechnological realm. Or we can be more sophisticated and ask of it why the arts of language that go into oral literature are any less of a

technology than those of the potter? The ability to perceive, from
the point of view of one's own language, possibilities in it for
rearrangement of elements ignored in ordinary discourse, and to
direct particular patterns of rearrangement to particular functions—this seems as much of a technology as the ability to deploy
marks on a surface to generate visually compelling representations, or the ability to construct and tune musical instruments.
There being nothing in nature before Edison to preserve actual
speech other than parrots and echoes (both of these preserving
only a few syllables, although one for long periods of time and the
other for very short ones), writing had to represent speech—even
as spoken language could be said to "represent," or signify, entities actions, processes, and so on—in as removed a way as drawing
represented what was seen. And I should add that, for purposes
of this discussion, it might also be stipulated that there are indeed
preliterary texts: neither the Hammurabic code nor inventories
in Linear B could be called prose, for example.

A few more preliminaries seem in order, though, on the concept of *technology* and what we might mean by it: we tend to think
of the progression of stone, pottery, bronze, wheel, iron, sail, oar,
ploughshare, yoke, clockwork, waterpower, steampower, industrial technologies multiplying like mad, then mechanical-electric-
electronic as perhaps a technological backbone extending along
the body of cultural history. (And whether we think of digital storage and processing as an extension of that or a new phase I am in
no position to discuss. Nor have I world enough and time to consider the matter of cinema and whether it is literature or rather
part of the matter of visual art.) But this paradigm tends to put
cruder hardware earlier, more complex hardware later, and move
toward softer and softer matters. Throughout, I shall be investi-
gating tools and techniques alike, and—given the way in which
they give rise to each other historically—to privilege neither.
There are indeed questions about how soft a technology can be,
and of literary texts in the broadest sense *as* technologies themselves,[1] that is, the vernacular Bible for the Reformation generally,

and the psalter in vernacular and the chorales and hymns growing from it in the sixteenth century as being pieces of liturgical technology for reformed churches of many sorts. Or, in the realm of visual imagery, both the emblem book as a piece of renaissance conceptual technology and the emergence of the diagram from allegorical picture (as Walter Ong [1958] pointed out in an important essay).

Like the double-entry bookkeeping and the rational musical notation that were analogous techniques for Max Weber, these all inhabit a realm so soft that some might want to reject the use of the term *technological* here as being perverse. The stage machinery of Inigo Jones allowed sets to change with almost magical effect in Stuart masques, but the masque itself as a piece of political machinery (as Stephen Orgel [1975] pointed out so well) some might want to call an institution instead. And this leads to the question of institutions, corporations, bureaucracies, governments, and so on, in general: do we think of their operations as *technological?* Or are typewriters and punched card systems and dictaphones (to mention some archaic ones) the technology, and the corporate structure represented by a flow-chart something else? I mention this because literary institutions—like poetic conventions, genres, forms (such as the adaptation of Greek meters by Latin poetry in the republican period, the epistolary form of the novel), interpretive communities, authorial roles, groups of readers, and so on—might then be adduced in favor of, declaring that literature *is* a technology, or cluster of technologies. But we shall only note these questions here, and observe that it would be relatively easy to rewrite the agenda of our topic by so saying and leave it at that.

But now to return to my central question, the one that perhaps accounts for why there is a subject here at all. It is the peculiar power of both art and literature to represent—to construct images, descriptions, metaphors, tokens, symbols, stand-ins, ad hoc and apragmatic conceptual schemes and fictional histories, sciences and philosophies, and a host of other artifacts with valuably problematic relations to the world and its parts. Technologies of various kinds can produce an endless array of models, as various as metaphors or fables. But it is only literature that can represent technology itself. Whether or not we arbitrarily consider it to be literature, it is modern historiography—economic, scientific, military, social—that comes up with the very notion. For English up through the eighteenth century, the analogous term would have been *art* (as opposed to nature); the phrase "the art of *x*" means, of course, "the craft or *technê* of *x*." And all this before the term *art* becomes elevated over the term *artifice,* which remains in the domain of what would later be called technology. In considering the recognition of technology in literature before the industrial age, it is important to remember that it is the notion of a particular skill (the Greek *technê*) that not only gives its name to our term, but which inheres in the Latin-rooted *art.*

A particularly poignant moment in our poetic history represents the aetiology of all our concerns here. Eden, in Milton's *Paradise Lost,* is not merely a place, but a condition that it would be wrong to call pretechnological, only because it is preternatural as well. There are no seasons (spring and fall are continuous, as on a lemon tree now), there is no cooking, no tools, no knowledge beyond phenomena, no institutions, and no death. The fall from this first condition of perfection can be considered a fall into the natural condition itself. At the end of Book IX, after Adam completes the original act by eating the fruit Eve has already tasted, something like the first earthquake and the first thunderstorm occur; following this, Adam and Eve turn to the first guilt-shadowed sexual act in human history, after which they sleep disturbedly, rise "as from unrest" (9.1051–2) and construe their former nudity as nakedness. It is significant that this is the occasion of the first or original technology. In avowing their *pudeur* at their nakedness, Adam suggests looking for

> Some Tree whose broad smooth Leaves together sew'd
> And girded on our loins, may cover round

Those middle parts, that this new comer, Shame,
There sit not, and reproach us as unclean.

(9.1095–8)

There had been perfectly innocent sex in Paradise before the Fall
into Eros and Thanatos, and into manual and obstetrical labor.
There had been nothing wrong with their having sex, but they displaced their guilt at disobedience onto lovemaking. But now, this
sexual shame leads Adam to talk casually of sewing, which thereby
comes into being as they find the fig tree (as described, actually a
sort of banyan):

Those leaves
They gather'd, broad as Amazoninan Targe,
And with what skill they had together sew'd
To gird their waist, vain Covering if to hide
Their guilt and dreaded shame; O how unlike
To that first naked Glory.

(9.1110–5)

Milton then interjects a revealing comparison:

Such of late
Columbus found th' American so girt
With feather 'd cincture, naked else and wild
Among the trees on Isle and woody Shores.

(9.1115–8)

They have suddenly become not the primary people, but rather
very primitive ones. Yet it is not that Adam and Eve revert to something prior, falling back down along a progressive sequence, but
rather falling—in another sense—into that sequence itself. It
might be objected here that there are indeed tools in Paradise:
Eve is said to work "with such Gard'ning tools as Art yet
rude/Guiltless of fire had form'd" (9.390–91). These unfallen,
pretechnological implements, employed in unfallen work (not

labor), again, in their prenatural status parallel the matter of unfallen sexuality. Both are matters that would seem to defy representation save in fallen terms, technological or erotic. In any event, it is clothing that is introduced as the first technology in the biblical story, and we should remember that the subsequent history of clothing—fabric, construction, disposition of semiological elements in it—remains, like the realm of cuisine, as significant a matter for fictional representation as the more sensational developments in rapid transportation and mass production of commodities. (For classical mythology, Arachne, in her contention with Athena as master-weaver, is as much a *faber* as Daedalus.) And it is important to remember that what Milton is relating involves no nature/culture distinction; rather, the two are reciprocal in the fallen condition. The only profound distinction between them might be that in nature there is death. Given this, all culture, all technologies both hard and soft, are prosthetic for loss of life.

What human technology might be like later on in history is ironically foreshadowed earlier in *Paradise Lost,* in the somewhat mock-heroic account of the War in Heaven, the failed revolt of Satan's rebel angels. In Book VI, at one point, the ad hoc invention of artillery is recounted. Significantly, the sequential relation of empirical observation to discovery to invention is elegantly implied in Satan's assertion that beneath surfaces lurk possibilities, remarking that, given

> This continent of spacious Heav'n, adorn'd
> With Plant, Fruit, Flow'r Ambrosia, Gems and Gold,
> Whose eye so superficially surveys
> These things, as not to mind from whence they grow
> Deep under ground, materials dark and crude,
> Of spiritous and fiery spume, till toucht
> With Heav'ns ray, and temper'd they shoot forth
> So beauteous, op'ning to the ambient light.

> (6.474–81)

This might be the Ur-science. Impatient with the acceptance of superficial phenomena, and acknowledging what lies behind (here, literally, *beneath*) them, a restless spirit of inquiry leads to invention, science to technology. On this basis, Raphael (who is telling this story to Adam in Paradise) observes that

> they [Satan's army] turn'd
> Wide the Celestial soil, and saw beneath
> Th'originals of Nature in their crude
> Conception; Sulphurous and Nitrous Foam
> They found, they mingl'd, and with subtle Art
> Concocted and adjusted they reduc'd
> To blackest grain, and into store convey'd;
> From hidden veins digg'd up (nor hath this Earth
> Entrails unlike) of Mineral and Stone,
> Whereof to found their Engines and their Balls
> Of missive ruin; part incentive reed
> Provide, pernicious with one touch to fire.

                                            (6.512–20)

And so the first Little Bang—but big enough—of destruction rather than origination, ensues.

Earlier in the poem (Books I and II), but later in the story, demonic technology—absent from the prelapsarian human world, but present in its Satanic prototypes—is represented by Mammon, whose crew "Op'n'd into the Hill a spacious wound/And digg'd out ribs of Gold." The eventual fallen human legacy of this is that men, thereafter, "Ransack'd the Center, and with impious hands/Rifl'd the bowels of thir mother Earth/For treasures better hid."[2] In the Greek account of human history, however, it is Prometheus who steals fire—and thereby, allegorically, all technology—from Olympus to give to men. *Technê* thus had, in another fable, its divine origins instead of, for Milton, its demonic ones. (And yet there is a common chord of resentment linking these stories: Prometheus is avenging what he conceives of as an unfairness to his brother Epimetheus, and Satan and his legions their loss of Heaven.)

Parallel to the biblical account of an unfallen condition, there is a classical one that Milton also had in mind. The ancient mythological progression downward from the golden age through silver, bronze, and iron ones has some reference to technology. Our modern archaeological scheme of the ages depends upon a technological sequence of progress, from neolithic to bronze to iron to steel. (And how would we continue? Steam, electric, electronic, microchipped?) Indeed, the old Persian and Greek Golden Age was a version of the biblical paradise, an unfallen state in which whatever was desirable was thereby good, in which there was no trace of sexual guilt, or shame, and in which justice—personified as the virgin Astraea—dwellt among men (she became stellified as the constellation *Virgo*). And perhaps her loss may be thought of as necessitating the engines of legal and judicial justice (if, again, institutions can be technologies). But Ovid, in his canonical account, makes it clear that there was no need for *homo faber*[3] in the Golden Age (I quote from Arthur Golding's lively translation of 1567):

> There was no fear of punishment, there was no threatening law
> In brazen tables nailed up to keep the folk in awe.
> There was no man to crouch or creep to Judge, with cap in hand:
> They lived safe without a Judge in every realm and land.
> The lofty pine tree was not hewn from mountains where it stood,
> In seeking strange and foreign lands, to rove upon the flood.
> Men knew no other countries yet than where themselves did keep;
> There was no town inclosed yet, with walls and ditches deep.
> No horn nor trumpet was in use, no sword or helmet worn:
> The world was such that soldier's help might easily be forborn.
> The fertile earth was yet was free, untouched of spade or

plough,
And yet it yielded of itself of everything enough.

The pine tree, content to stay at home rather than becoming the tall mast of a ship, is a charming central image here, but the commingling of tools, social roles, and institutions here is interesting as well.

We might now consider the matter for literature of the imaginative energy arising from the contemplation of the particular products and procedures of the individual artifex and his or her craft. Wonder at mechanical skill, sometimes associated with magic, is an old matter. But from late antiquity through the Renaissance, crafted objects have projected iconographic adumbrations, perhaps by association with allegorical personifications: sword, scales, scythe, hammer, stringed instrument, mirror (to name just a few); this last serving also as a fundamental metaphor for instructive representation, and for the exemplary, in the fifteenth and sixteenth centuries (Ophelia calls Hamlet "The glass of fashion"; a set of cautionary tales of the downfall of celebrated people first published in 1559 is called *A Mirror for Magistrates* and George Gascoigne's satirical poem of 1576 is called *The Steele Glass,* to indicate that its picture of the world will hardly be flattering).

The material for a lot of poetry thus lay in further interpretation, by glossing or by other use, of crafted objects and utensils. One instance is in the so-called metaphysical or baroque conceit—a complex metaphor worked out in often elaborate detail—characteristic of much lyric poetry of the Renaissance. Shakespeare is always developing ad hoc conceits, in the speech of his characters, which draw on the wisest range of human arts and institutions. One of the most notorious and best known metaphysical conceits—of the sort that Samuel Johnson famously condemned as exhibiting "dissimilar ideas yoked by violence together"—is Donne's drawing-compass metaphor in "A Valediction: Forbidding Mourning." This is a poem of consolation for a temporary parting, and its strategy is to reinterpret separation in a more hopeful way. He introduces his pointed trope (of what is indeed a pointed implement) by way of another introductory but unrelated mechanical image, that of the remarkable spread of gold leaf as it is made. This is followed by several applications of the compass

> Our two souls, therefore, which are one
> Though I must go, endure not yet
> A breach, but an expansion
> Like gold to aery thinness beast.
>
> If they be two, they are two so
> As stiff twin compasses are two,
> Thy soul the fixed foot, makes no show
> To move, but cloth, if th' other do.
>
> And though it in the centre sit,
> Yet when the other far cloth roam,
> It leans, and hearkens after it,
> And grows erect as it comes home.
>
> Such wilt thou be to me, who must
> Like th' other foot, obliquely run;
> Thy firmness makes my circle just,
> And makes me end where I begun.

Sometimes the revisionary force of substituting contemporary for earlier technology seems to acknowledge some of the self-consciousness of literature's rhetorical uses of contemporary technology: Donne in "The Broken Heart" uses chain-shot as a substitute for Love's traditional bow-and-arrow, in an observation of how Eros mows down whole ranks, a military conqueror rather than just a cute, blindfolded boy out hunting for what he can get. Later in the seventeenth century, Robert Herrick and Ben Jonson both adapt a neo-Latin poem on an hourglass in nicely differing ways. In Jonson:

Do but consider this small dust
Here running in the glass,
By atoms moved;
Could you believe, that this,
The body was
Of one that loved?
And in his mistress' flame, playing like a fly,
Turned to cinders by her eye?
Yes; and in death as life unblessed,
To have't expressed,
Even ashes of lovers find no rest.

("The Hour-Glass")

Herrick's poem—from the same Latin original, but also obviously ("as I have read") knowing Jonson's, which was published eight years earlier (1640), and perhaps circulated by 1619—ingeniously substitutes drops of water for grains of ash, thereby perhaps combining hourglass and Greek clepsydra:

That Houre-glasses, which there ye see,
With Water fill'd (Sirs, credit me)
The humor was (as I have read)
But Lovers tears enchristalled.
Which as they drop by drop do passe
From th' upper to the under-glasse,
Do in a trickling manner tell,
(By many a watrie syllable)
That Lovers tears in life-time shed,
Do restless run when they are dead.

("The Houre-glasse")

Not all treatments of mechanical artifacts are emblematic in conventional or ad hoc ways, of course. A wonderful anonymous fifteenth-century English poem seems to proceed from direct experience rather than modulated by an inherited interpretive commonplace. It inveighs against the noise of a smithy (I have modernized the late middle-English):

> Black-smocked smiths, smattered with smoke,
> Drive me to death with din of their dints!
> Such noise at night ne heard men never:
> What knavish cry and clattering of knocks!
> The crooked cowards cry out "Col, col!"
> And blow their bellows till their brains burst.
> "Huff, puff!" says that one, "Hoff-poff" the other. . .
> Of a bull's hide are their big leather aprons,
> Their calves are guarded against fiery sparks.
> Heavy hammers they have that are handled hard,
> Sharp strokes they strike on an anvil of steel.
> "Bang, bang! Lash, dash!" go their answering crashes:
> So doleful a dream let the Devil dispel!
> The boss takes a big piece of iron and binds it
> To a tiny one, bangs it, and twangs out a treble.
> Tick tock! Hick, hock! Tickit, tockit! Tick, tock!
> Bang, bang! Lash, dash!—such a life they lead!
> May Christ give all horse-shoer plenty of sorrow;
> For these water-sizzlers no man at night has his rest
>
> (Arundel MS.292)

This contrasts strongly with the allegorical smithy of Care in Book IV (Canto V, st.32ff.) of Spenser's *The Faerie Queene*, about 150 years later, in which that blacksmith and his six servants "to small purpose yron wedges made;/These be unqiet thoughts that care-ful! minces invade." The hammers of all seven of these smiths were of various sizes and, "Like belles in greatnesse [did] orderly succeed"—by this simile Spenser invokes an old story that Pythagoras had discovered the relation between musical intervals and the relative sizes of vibrating bodies by hearing different pitches coming from different sized bars of iron in a forge. To fig-ure individual anxieties as small metal wedges made to no partic-ular purpose is a brilliant move. But the later-medieval lyric poem—save perhaps for a trace of this in the "treble" produced by a small piece of iron—seems all from direct observation and with no allegorical concerns at all.

Another question is that of the consideration in poetry and fiction of machines whose function was to yield knowledge, rather than to do physical work. We have already considered emblematic treatments of the hourglass—one of those machines, in fact—considered as a homely object. An easily manageable example of what seems already to be acknowledged technology in our sense might be that of the telescope as dealt with in seventeenth-century poetry. It was a prominent and even sensational new invention, particularly in its avowed role of helping to contrive the "new philosophy"—the Copernican and Galilean revisions of the way things are. Milton in *Paradise Lost,* Book I, doesn't notice the implications of a consequent Copernican cosmology, but rather ingeniously associates what is seen through the telescope with a realm of illusion and wonder. He describes Satan's shield, whose "broad circumference"

> Hung on his shoulders like the Moon, whose Orb
> Through Optic Glass the *Tuscan* artist views
> At Ev'ning, from the top of *Fesole*
> Or in *Valdarno,* to descry new Lands,
> Rivers or Mountains on her spotty Globe.

The simile is simply one of scale, but the qualification—the moon as it loomed up hugely for Galileo—calls into question the basis for the wonder, and the tainted epistemology of getting beyond phenomena.

On the other hand, William Davenant (in 1650, before *Paradise Lost*) considers the telescope instrumental to a quest for truth, and even a moral force:

> Others with Optick Tubes the Moons scant face
> (vaste Tubes, which like long Cedars mounted lie)

Attract through Glasses to so near a space,
As if they came not to survey, but prie . . .

Man's pride (grown to Religion) he abates,
By moving our lov'd Earth; which we think fix'd;
Think all to it, and it to none relates;
With others motion scorn to have it mix'd;

As if 'twere great and stately to stand still
Whilst other Orbes dance on; or else think all
Those vaste bright Globes (to shew God's needles skill)
Were made but to attend our little ball.

(Gondibert II.5)

During the later seventeenth and eighteenth centuries, land-scape gardening—a subject for topographical verse—and other georgic poetry will focus on agricultural arts. But it is of course with steam-driven industrial technology—when the word *mill* starts to designate a factory and the industrial revolution starts to proliferate new machines designed to produce new artifacts, and with processes and hardware demanding developments in each other in a pattern related to the historical dialectic on invention and discovery—that technology in our general sense can become an issue for imaginative literature. (I have been exempting histo-riography considered as a literary genre all along only because, not being a historian, I can't comment on the way historical writ-ers from Herodotus on were variously aware of technological change and variously interpretative of its wider consequences. Similarly, I have contracted not to deal with the histories either of philosophy or of social thought, although literature has recently laid claim to both.) The industrial revolution could even be said to effect linguistic changes—for example, the proliferation of hyphenated compound nouns resulted from the proliferation of new products in the world. They required new names, and unless the invention were from abroad and came with a ready-to-adapt-or-translate French name, the new entity would be named by a compound, hyphenated at first (for example, *steam-boat*).

But deliberately negative considerations of utensils and processes could begin to arise in the eighteenth century. William Cowper, in a poem about a field of poplars being cut down, observes with a mixture of ruefulness and reconciliation that "the tree is my seat that once lent me a shade." William Blake's elaborate vision distinguished between the mythological fabrication of the Vulcan-like Los in his long poems, and the actualities of industrial England. Blake's "dark, Satanic mills," no matter how they were subsequently misconstrued, were not literal factories so much as allegorical engines of compression and reductiveness of the kind founded upon constrained repetition ("the same dull round, even of a universe, would soon become a mill with complicated wheels" [*There Is No Natural Religion*]). And yet, consider these lines from *Jerusalem* (1804):

> I turn my eyes to the Schools and Universities of Europe
> And there behold the Loom of Locke whose Woof rages dire
> Washd by the Water-wheels of Newton. Black the cloth
> In heavy wreathes folds over evry Nation; cruel Works
> Of many Wheels I view, wheel without wheel, with cogs tyrranic
> Moving by compulsion each other: not as those in Eden: which
> Wheel within Wheel in freedom revolve in harmony and peace
> (Plate 15)

But a century later, after some acquaintance with the celebration of the machine in Futurist art and literature as well as following a direction taken by Whitman earlier, Hart Crane, in the "Cape Hatteras" section of *The Bridge* (1930), responds to the machinery—belt-drive, in particular—in a way quite opposite to Blake's parable of geared wheels as exerting dreadful power over one another. Crane praises the energy of dynamos, taking the generation of electrical power as a metaphor for imaginative power. He hears "new verities" in the humming of rotors, even as he rejoices in reading how "Power's cript,—wound, bobbin-bound, refined—/Is stropped to the slap of belts on booming spools." And later on in the same passage, even the conditions of restraint ("axle-bound, confined/In coiled precision") the turning wheels rejoice in their spinning.

But—to return for a moment to Blake's emblem of gears—we can note another matter. The Edenic "wheel within wheel" is an adaptation of the concentric wheels of Ezekiel's vision, but as Blake describes them, they seem to be transmitting power outward. A contemporary reader will see this as a kind of fluid drive used in automatic automobile transmissions. Frequently, a purely visionary construction in poetry will seem to anticipate a mechanical or other invention of a much later time. The implicit trope of deep freeze or even cryonic preservation in the "cold pastoral," the frozen activity, on the band of images around Keats' Grecian urn; the cinematic montage that Eisenstein found implicit in sequences of simile in *Paradise Lost*. Spenser has an image—in a lament for Sir Philip Sidney—of the dead poet's harp becoming stellified (as the constellation *Lyra*) and, as it moves upward, having the wind play across its strings. An eighteenth-century critic of Spenser sees this as an unwitting invention of the Aeolian harp, an adult toy of purely eighteenth-century invention. The history of poetic fiction is full of these instances, which are mostly amusing accidents. But one has only to think of more general mythical activities and metaphors whose consequences for our conceptualization of the world are more far-ranging. There is human flight—which has become literal to the point of being scheduled—or, as Rosalind Williams has so elegantly discussed in her *Notes on the Underground* (1990), the literalization of a previous figuration through technological development.[4] Wordsworth, in a late sonnet entitled "Steamboats, Viaducts, and Railways" (1833) forthrightly avers that even his own celebrated view of the supremacy of nature must yield to mental—and instrumental—power, and "embrace/Her lawful offspring in man's art":

> Motions and Means, on land and sea at war
> With old poetic feeling, not for this,

Shall ye, by Poets even, be judged amiss!
Nor shall your presence, howsoe'er it mar
The loveliness of Nature, prove a bar
To the Mind's gaining that prophetic sense
Of future change, that point of vision, whence
May be discovered what in soul ye are.
In spite of all that beauty may disown
In your harsh features, Nature cloth embrace
Her lawful offspring in Man's art; and Time
Pleased with you triumphs o'er his brother Space,
Accepts from your bold hands the proffered crown
Of hope, and smiles on you with cheer sublime.

But what about the artifacts and "means and motions" of technology as bastards of nature? The notion surfaces much earlier, in a well-known scene in Shakespeare's *The Winter's Tale* about flowers interbred by grafting, producing new varieties "which some"—says Perdita—"call Nature's bastards."[5] Over time, the legitimacy of some of these becomes differently perceived. Older technologies get taken for granted, and subsumed somehow into the natural world. An old stone mill with a half-ruined wooden overshot wheel looks like part of a picturesque scene; and what about an old horse-drawn hayrake or disc-harrow? Not yet a trashed car body, though.[6] If one mode of pastoral realm seen as utopian for an industrial world might be one in which there were no trade-names but only generic ones, then an analogue of this might be a name for an object that is no longer known to be, or derive from, a trade-name (for example, the lost matter of something called Rowland's Macassar Oil in the name of an "anti-macassar"). In literature, this historically changing perception or construction of tools and techniques can be observed in, for example, the tendency to subsume under the term *pastoral* all matters merely rural; specifically, the kind of poetry coming from Hesiod's *Works and Days* and Virgil's *Georgics* which is devoted not to what was always a mythological realm, but to the technologies of farming. It is as if agricultural work hadn't employed the products of industrial manufacture for centuries.

When, a half-century after Wordsworth's sonnet, the American poet Christopher Pearse Cranch can write a sequence of sonnets on *Seven Wonders of the World,* his poems "The Printing Press," "The Locomotive," and "The Photograph" all contrast these modern wonders with those from classical mythology, rather than with the traditional seven wonders of the ancient world, most of them the products of the most sophisticated technology of their times. Thus, his sonnet on the railroad engine:

> Whirling along its living freight, it came,
> Hot, panting, fierce, yet docile to command —
> The roaring monster, blazing through the land
> Athwart the night, with crest of smoke and flame;
> Like those weird bulls Medea learned to tame
> By sorcery, yoked to plough the Colchian strand
> In forced obedience under Jason's hand.
> Yet modern skill outstripped this antique fame,
> When over our plains and through the rocky bar
> Of hills, it pushed its ever-lengthening line
> Of iron roads, with gain far more divine
> Than when the daring Argonauts from far
> Came for the golden fleece, which like a star
> Hung clouded in the dragon-guarded shrine.

On the other hand, one might see another sort of poetic wonder in the direct acceptance of the railroad and a consequent energetic attention to observable phenomena attending upon its use. These lines of Dante Gabriel Rossetti's are from a long verse-letter called "A Trip to Paris and Belgium" (1849); it may be the first poem in English to consider what the world looks like through the window of a moving train:

> The country swims with motion. Time itself
> Is constantly beside us and perceived.

Our speed is such the sparks our engine leaves
Are burning after the whole train has passed.

The darkness is a tumult. We tear on,
The roll behind us and the cry before,
Constantly, in a roll of intense speed
And thunder. Any other sound is known
Merely by sight . . .
Our speed has set the wind against us. Now
Our engine's heat is fiercer, and flings up
Great glares alongside. Wind and steam and speed
And clamor and the night. We are in Ghent.

                                        (sect. XVI)

The moving train in nature merits all the close observation a
botanizing observer, for example, might give it. The engine is
acknowledged in a deeper way, perhaps, than by prophetic procla-
mations, such as that of Walt Whitman in *Passage to India,* published a few years before Cranch's series, and "Singing" as he said
"the great achievements of the present":

Singing the strong light works of engineers,
Our modern wonders (the antique ponderous Seven outlived,)
In the Old World the east the Suez canal,
The New by its mighty railroad spanned,
The seas inlaid with eloquent wires . . .

                                        (3–7)

This is imaginative exuberance in behalf of what seems to need
no redemption. Another mode of allegorical appropriation
can be adduced that raises questions about attitudes toward
modernity and novelty in a technologically stagnant community. I
cite one example of an epigrammatically homiletic sort here
(from Martin Buber's *Tales of the Hasidim):*

"You can learn something from everything," the rabbi of Sadagora said to his hasidim. "Everything can teach you something, and not only everything God has created. What man has made has also something to teach us."

"What can we learn from a train?" one hasid asked dubiously.

"That because of one second one can miss everything."

"And from the telegraph?"

"That every word is counted and charged."

"And the telephone?"

"That what one says here is heard there" (Buber, 1948, p. 70).

With never a trace of a redemptive allegorical agenda in mind, however, Ezra Pound's Hugh Selwyn Mauberley, the esthete born "in a half-savage country, out of date" complains that "The pianola 'replaces'/Sappho's barbiton" with sarcastic quotation marks about the verb.[7] But in a memorable scene from Hermann Hesse's *Steppenwolf,* Mozart brought back from the grave delights in saxophones and the jazzy music they are playing. Even if irritation at newfangledness is backed by some larger sense of history, it is as open to satiric attack as the novelties at which it shudders at.[8]

Of great interest are the ways in which technological objects and procedures are represented in two differing modes of prose fiction—the relatively realistic novel and the romance that preceded it (and, in the nineteenth century, lived an interestingly marginalized life, in horror story, literature for children, and American books like *The Scarlet Letter* and *Moby-Dick.* In romance, technologies are visionary or notional: magic, illusion, transformation, allusive relations among places, stories, individual persons, and their prototypes in previous literature. The novel becomes concerned with—alongside its broadening concern with social structure, character, mobility and so forth—the role and presence of technological development and change.[9] The minutiae of the crafts, industry, and business of printing explored in one part of Balzac's *Les Illusions perdues* might be compared with the total immersion of *Moby-Dick* in the matter of whaling. In Melville's case, both larger and smaller components of a complex technology come up for emblematic and allegorical consideration, as if major portions of the book were elaborate secular sermons preached on the text of the hard facts. But these are the devices and strategies of poetry, as we have seen, rather than what we come to think of as novels, whose concerns—I mention only a few: Dickens' *Hard Times,* Zola's *Germinal,* Frank Norris' *The Octopus*—are more with the effects of technology on human lives and institutions. It continued for poetry and for meditative essay to continue to consider evolving technology with wonder, distaste, appreciation, fear, ironic appraisal, various ambivalences, and a host of other attitudes generated uniquely by any particular text. And all of these attitudes may be framed in more manifest or latent ways, and variously construing as more natural or more mechanical the concrete facts of life beyond our bodies.

Utopian fictions, tending to take on the form of romance, tended before the middle of the nineteenth century to come up with occasional imaginary, technological inventions; what we call science fiction is of course a mode of romance, sometimes utopian, usually dystopian. (At its least interesting it is neither, merely extrapolating from the twentieth-century history of military and transportational technologies to continue what were essentially Renaissance epics of conquest and combat.) The prophetic invention of atomic energy in H. G. Wells' remarkable novel *Tono-Bungay* is an ancillary matter in its unfolding story (which even more remarkably and originally deals with he world of advertising): the book is not of that genre at all. Science fiction perhaps should be thought of as technological romance, centered more on situations and predicaments than on character and relationship.

The mechanical or other sorts of simulation of life is an enduring matter for romance. Such projects are always felt to be fraught with dangers—as if blaspheming the creation of mankind in Genesis by implicitly considering even that to be technological. A brief chronicle of the history of such fictions might include the following: the late medieval legend of the Golem; Spenser's stories of simulacra in Books I, II and V of *The Faerie Queene* (Talus, the iron man—an anticipatory robo-cop; Archimago's false Una; and, even more remarkable, the False Florimell, a beautiful creature made of snow and animated by a male demon); the animated, life-sized doll of E. T. A. Hoffmann's Dr. Coppelius, the created figures in Frankenstein, Hawthorne's uncanny Feathertop; beneficent mechanical personages like the Tin Woodman and Tik-Tok in L. Frank Baum's *Oz* books, and—more in a Hawthornian and Spenserian mold, the central figure of Thomas Pynchon's *V.*

The form of romance has often been used by satire, *Gulliver's Travels* being a prime example. That book's various angles of vision are focused, in the Lilliput/Brobdingnag sections, through grotesquely great changes of scale, onto the technological particulars—as well as, more obviously, the social, moral, and political norms—of early eighteenth-century England. These are fully as interesting in their reflections of decades of accepted tele- and microscopy, as is the more obviously proto-science fiction in the account of Laputa, with its flying island world and its devastating satire of the Royal Society. (Swift uses fantastic technology itself as a satiric device against philosophical over-literalizing, particularly in the digression "On the Mechanical Operation of the Spirit" from *The Tale of a Tub,* although mechanical contrivances are frequently bracketed as comical from Rabelais on.)[10] But in general, the concept of the mechanical as opposed to the organic becomes almost a cliche during the nineteenth century.

Imaginative ambivalence about the certainty of machinery and the uncertainty of human and other organisms provides an amusing moment in Poe's essay on the chess-playing automaton of Baron von Kempelen, but displayed in the name of Johann Nepomuk Maelzel (inventor of the metronome among other devices), that was exhibited in the United States after its inventor's death.

It is a piece of investigative reporting, and cleverly demonstrates how—despite all sorts of deceptive demonstrations by the operator that the chess-player is pure machinery—there is a midget chess-master hidden inside it. (He was correct.) But his deduction starts with a charmingly wrong premise: although the machine almost always won (when playing what one could imagine were the patzers who came in Baltimore to take it on), it had been known, although very rarely, to lose a game. From this Poe concluded that it could not be a machine, because a machine would have had to win all the time: here, the opposition of the organic and the mechanical is linked with that of imperfection and perfection. I remember telling this story to Claude Shannon around 1955 or so, when he had finally devised a chess-playing program for then very slow and bulky state-of-the-art computers. He said that it always lost, but that it was very gradually getting better. This, of course, seemed as truly spooky at the time as it seemed, after all, inevitable.

I shall conclude with the mention of one last work of fiction. William Golding's *The Inheritors* is a major piece of mid-twentieth century romance (and for me a finer book than his previous *The Lord of the Flies*). Its narrative follows in painstaking detail, as they slowly migrate from their summer to their winter quarters, a group of very primitive *homo sapiens*—tool users but not tool keepers, for everything they fashion is ad hoc and cast aside after use, like the tools of some birds and animals. The reader is beautifully trapped into feeling that we are the descendants of these simple but benign people, until, in a climactic moment in the book, they encounter a different group of men who are clearly revealed to be a different species of *homo faber*, possessed of technology and property and enterprise and aggressive homicide. And it becomes shockingly apparent that the people we have been attentively and sympathetically following throughout the book will succumb to these, whose inheritors in fact we are.

Notes

1 We might note here what Leo Marx reminds us of elsewhere in this issue about the original use of a *technology* to mean a manual or methodology, then the study of or knowledge about practical arts. One instructive early use (*OED* gives 1704) of *technological* is in "a technological term," where we would say a "technical" one.

2 2.686–88; And here he adapts material from Ovid in his account of the age of iron.

3 And see Arendt, 1959, p. 341, on the unknown source of this phrase, and the relations among the terms *artifex, opifex* and *faber* (equivalent to Greek *tekton*).

4 Such chains of object and literary metaphor—as in the history of actual and mythical gardens and their interrelations throughout history, for example—is a complex question. Literary fictions can allude to and deal with and otherwise contain previous ones; new technologies frequently merely supersede earlier ones. A historian of technology might be able, perhaps, to point out analogues of literary revisionism in a sort of "pseudoconsciousness" of as prior invention in a later one.

5 The debate about the naturalness or artificiality of grafting continued in the seventeenth century. See, for example, Andrew Marvell's "The Mower Against Gardens" for its witty interweaving of technology and sexuality.

6 A recently translated essay by Jean Starobinski entitled "Water-Wheels" deals elegantly with some of these questions, and with what he refers to as "the poetic or narrative representation of a *visible* aspect of industrialization." And again, I must cite here the work of Leo Marx, generally.

7 And see Elizabeth Eisenstein's observations on "supercession" in her remarks in this issue.

8 I should add that the dense history of ambivalences about technology in American literature of the nineteenth century has been so authoritatively studied by Leo Marx in *The Machine in The Garden* that its explorations and arguments alone could have served as a discussion of out present topic.

9 Early on in the history of the novel, *Robinson Crusoe* might be mentioned as chronicling the loss and then the improvisatory reconstruction of the ordinarily invisible or unregarded technologies required for daily existence.

10 For example, the ingenious wheel-mechanism for equally distributing the sound and aroma of farts in Chaucer's "Summoner's Tale" to the

splendid contraptions of Rube Goldberg, which satirize the relation of mechanism to function, among other things.

References

Arendt, Hannah, *The Human Condition* (New York, Anchor Books, 1959).

Buber, Martin, *Tales of the Hasidim, The Later Masters* (New York: Farrar, Straus and Young, 1948).

Havelock, Eric A., *The Muse Learns to Write* (New Haven, Yale University Press, 1986).

Ong, Walter J., "From Allegory to Diagram in the Renaissance Mind," *Journal of Aesthetics and Art Criticism* 17 (1958): 423–40.

Orgel, Stephen, *The Illusion of Power* (Berkeley and Los Angeles: University of California Press, 1975).

Starobinsky, Jean, "Water-Wheels," *The Hudson Review* 49:4 (Winter, 1997): 553–68.

Svenbro, Jesper, *Phrasikleia: An Anthropology of Reading in Ancient Greece* (Ithaca: Cornell University Press, 1993).

Williams, Rosalind, *Notes on the Underground* (Cambridge: MIT Press, 1990).

The Arrival of the Machine: Modernist Art in Europe, 1910–25

BY ROBERT L. HERBERT

I APPROACH the subject of art and technology rather like one of the blind men in the old story about the discovery of the elephant. The whole beast is too large to encompass at once, yet if I put my hand on the trunk I won't be describing the leg, if on the leg, then I won't be describing the head. As a historian more comfortable with concrete cases than broad generalizations, I'm going to investigate one portion of this elephantine subject, trusting that this will be an adequate sample of its complexity. I'll limit myself to the years bracketing World War I, when modernist painting and sculpture first paid wide attention to modern machinery, science, and industry.[1]

I write "modernist" and not "modern" for of course when we look outside the modernist canon to the whole range of visual culture—illustrations in the press, prints, photographs, film—we will find lots of images of modern industrial and scientific forms. These would constitute a rich body of material for a study of art and technology but I am going to set them aside as a future project and write here about some of the art that has become central to our cultural constructions by virtue of its dominance in public exhibitions, museums, the press, and in the literature of art history. In doing this I'm aware that it's not really the past we're talking about when we study the history of modernist art but our present culture, its desires, its fears, its expectations. Another limitation: I won't deal with the material side of art (the changing sources of its pigments, for example), although a study of this realm would show the imbrication of the visual arts in industrial techniques and products.

Before 1910

Why was it that avant-garde art kept the machine at arm's length until about 1910? In the two preceding decades, most figures and settings of modernist art were far removed from the modern city and its industrial forms. Various kinds of "primitivism" became virtually synonymous with modernism, ranging from techniques in painting, sculpture, and print-making to images of peasants and non-European peoples. Some artists, most notably Gauguin, sought out "primitive" places far from the modern city. Advanced taste at the turn of the century celebrated Cézanne's Provençal landscapes, Van Gogh's and Pissarro's peasants, Rodin's and Renoir's nudes, Gauguin's Tahitians, Signac's Mediterranean ports, and Monet's waterlilies. For, as the industrial revolution had taken an ever-firmer hold on culture in the nineteenth century, modernist art had become increasingly associated with premodern or primitive nature. Monet's famous paintings of the Gare Saint-Lazare in 1877 do not argue with this observation, because he never again painted any machines or, indeed, any industrial setting. His next fifty years (born 1841, he died in 1926) were devoted to non-Parisian scenes, climaxing in hundreds of pictures of his water gardens, which, in our end-of-century culture, are the most sought-after objects for blockbuster exhibitions.[2]

The extraordinary popularity of Monet's waterlilies in our high-tech era can remind us that art and nature are not fixed realities but constantly changing mental constructions. Often intertwined in the nineteenth century, they embodied release from the tensions of the workplace. They were compensatory domains, hence the dramatic rise of landscape and paintings of rural life as the industrial revolution took hold. In technique and style, a parallel to this is found in modernists' love of handcraft in opposition to the uniformity of academic art, which they likened to machine production. Instinct and spontaneity, considered natural, rather than machine-age regularity, characterized early modernist arts and crafts and their championing of individualism. Their sign was the irregularity of handwork, which Thorstein Veblen identified as the "honorific mark" that conferred value.[3]

As the most obvious materialization of modern industry, the machine only became important in advanced painting in the half-decade before World War I. I will shortly demonstrate that this was so by pointing to leading examples, but first I want to puzzle over the reasons for this sudden adoption after long delay, one of the fundamental mysteries of modernist painting and sculpture. Of course it is involved in the across-the-board innovations of Cubism, Italian Futurism, Russian Cubo-Futurism, and British Vorticism, but merely to say so isn't to explain why machine imagery suddenly loomed so large. Until about 1910, the machine was the enemy of all that modernist artists held dear: handcraft, creativity, individuality, and the marks of original expression. It bore the taint of mass-produced polish, an uncreative "perfection" that reduced the worker to a mere robot, and was also tainted by association with the money-grubbing bourgeoisie rather than with the free-spirited and marginalized artist.

If the machine were to be accepted by the vanguard, indeed to be taken as a leading sign of modernity, then it could no longer be considered the enemy. What permitted this *volte-face*? One factor was that, by the end of the century, the heirs of the international Arts and Crafts movement had accepted some industrial methods of production. The principle of the primacy of handcraft could appear to be maintained while in fact it was being rapidly undermined. For example, the art press increasingly used mass-production methods and photographic "process prints" to reproduce the effects of handcrafted images. When industry could be treated as the servant of creativity, there was no longer such a gap between handwork and the machine. Symptomatic of this change was the dedication of the Deutscher Werkbund in 1907 (Peter Behrens, Hermann Muthesius, Henry van de Velde, and others) to "the ennoblement of handiwork through the union of art, industry and handcraft" (Forgács, 1995, p. 6).

Another possible factor in accepting the machine as part of modernity rather than its opponent was the rapid increase of machinery in daily life. Bicycles and automobiles were commonly seen in the first years of the century, so were domestic sewing machines and electric lighting. Cheaper means of photographic reproduction allowed advertisers in the daily and periodic press to increase pictures of domestic appliances and machines, while the cinema became the newest form of mechanized entertainment. In offices, there were now typewriters, pneumatic tubes, and dictating machines. In the outdoor environment, one saw subways, elevated trains, trams, and seaport cranes. All these were quite new or a vast extension of preceding inventions, and so the machine became a normal, if not a fully domesticated, feature of daily urban life. By contrast, the industrial engines, locomotives and steam tractors of the nineteenth century were more external, more like monstrous invaders that could not be domesticated.

Of course the familiarity of the machine in daily life didn't mean that it appeared in the work of all modernists, nor that it was an indispensable aspect of modernism. Picasso's and Braque's paintings and collages (glued-together newsprint and other fragments) evoke café and studio still lifes or nudes: "fine arts" subjects rather than machinery. Modernity is found in their new language of fragmentation as well as in the evocations of contemporary intellectual life. The hegemony of their art is so great that Cubism is identified with them, and historians therefore pay little attention to the Epic Cubists[4] who identified modernity with the segmented imagery of such structures as locomotives, iron girders, and airplanes.

Typical of Epic Cubism is Albert Gleizes' *Landscape* (1914, Yale University Art Gallery), centered upon a city of excitable units of geometric architecture and scattered iron girders. Nearly abstract but recognizable pictographic forms identify railroad signals and a locomotive charging towards a river bridge while above the angular city, factory chimneys and smoke mark the nearby suburbs.[5] Jacques Villon made three paintings of factory machinery interpreted in the dynamics of pictorial structure,[6] and his brother Raymond Duchamp-Villon sculpted a *Horse* (Museum of Modern Art, and other collections) that combined machine parts with the organic rhythms of that animal. (The third brother, Marcel Duchamp, also investigated the look and meaning of machinery, as we shall see when we turn to Dada art.)

Another Epic Cubist, Robert Delaunay, dealt with the modern city and modern engineering. In a number of paintings from 1910 onward, he featured the Eiffel Tower of 1889 as the leading symbol of the new Paris. It appears in the upper right corner of his *Homage to Blériot* of 1914 (figure 1), a modern history painting and a masterpiece of Epic Cubism that summons up the famous aerial crossing of the English Channel in 1909. Above are two airplanes, while in the lower left, intertwined with other forms, is the large propeller and undercarriage of another airplane. Dominating the whole canvas are swirling disks whose partitioned colors call up the moon (cool colors) and the gaseous fireball of the sun (hot colors), forms of celestial dynamism linked with the machines that let humans overcome gravity. Delaunay hoped to engage the viewer in his conception of simultaneity, in which perception, stimulated by the optics of painted color, would bring awareness of the transcendent ideas of expansion and movement. Both form and subject would reveal the pervasive universal flux in which modern machinery is situated.[7]

Simultaneity was defined differently by the Italian Futurists, although they also linked it with the modern city and machinery. For them it was not as rarified a term as Delaunay's but instead a concept of immediacy, in which implied movement involves the

FIGURE 1. Robert Delaunay, *Homage to Blériot*, 1914. Oil, 250.5 x 251.5 cm. Basel, Oeffentliche Kunstsammlung Basel, Kunstmuseum.

spectator in an active engagement with the politics of modernity. The Futurists proclaimed modern technology as the instrument that would wrench their country violently into the modern world. In *What the Street Car Told Me* (1910–11, on loan to the Städtische Galerie, Frankfurt), Carlo Carrà uses vivid diagonals and tumbling geometry to suggest the interpenetration of a careening bus with nighttime lights, reflections, and passersby. It's fully as abstract as the work of Gleizes or Delaunay but has a visceral immediacy foreign to their more aloof calculations.

It wasn't pictures of machines but pictorial dynamism that expressed the new industrial age. Even less representational than

Carrà's canvas is Giacomo Balla's *Speeding Automobile* (1912, Museum of Modern Art, New York). Here, and in several variations on this theme, Balla embodied machine movement in a nearly abstract geometric vivacity partly inspired by motion photography. Balla seems to give visual form to some of the famous sentences from F. T. Marinetti's "Futurist Manifesto" of 1909:

> We affirm that the world's magnificence has been enriched by a new beauty: the beauty of speed. A racing car whose hood is adorned with great pipes, like serpents of explosive breath—a roaring car that seems to ride on grapeshot is more beautiful than the *Victory of Samothrace*.
>
> We will glorify war—the world's only hygiene—militarism, patriotism, the destructive gesture of freedom-bringers, beautiful ideas worth dying for, and scorn for women (Marinetti, 1909, pp. 21–22).

The latter phrase, of course, is one of the most flagrant proofs of the links between male sexism and violence, well hidden in Cubism but overt in Futurism.[8] In 1911 and 1912, the Futurists were ardent supporters of Italy's war with Turkey and the use of aerial bombardment in Libya. Then, after the onset of World War I in August, 1914, they did everything to help push Italy into the conflict on the allied side, publishing prowar pamphlets, organizing meetings, and the like. When Italy joined the allies in May, 1915, the Futurist leader Marinetti mustered most of the group into a military motorcycle corps that he organized. Subsequently he and several other Futurists joined Mussolini's new Fascist order and in the 1920s became propagandists for rapid industrialization. Was it not Mussolini who called Marinetti "the Saint John the Baptist of the Fascist movement?"

Elsewhere, before the war, the machine was associated with radical modernism. In England in 1913, Jacob Epstein sculpted his original *Rock Drill* (later dismembered) by incorporating pieces of a real drill atop a metal tripod. Later he cast in bronze the uppermost portion, a geometrized torso surmounted by a head in the shape of a welder's helmet.[9] In British Vorticism (Cork, 1976),

Wyndham Lewis and his colleagues on the eve of war proclaimed in masculinist terms echoing Futurism that modernity meant violent upheaval. Lewis's prewar *Plan of War* of 1914 (lost, known from a photograph) is abstract, but full of slashing geometry loosely related to diagrams of battle manoeuvres. His elaborated drawing *Combat No. 2* (Victoria and Albert Museum) of the same year, also a prewar work, shows mechanomorphic men fighting, their forms having the dynamic geometry of his abstract constructions. With the onset of war, he and other Vorticists specialized in pictures that indulge in the mechanisms of wartime violence, for example, Christopher Nevinson's *A Bursting Shell* of 1915 (London, Tate Gallery) or Edward Wadsworth's *War Engine* of the same year (now lost). The latter, although abstract, includes elements that were clearly patterned on pistons of an internal combustion engine.

Among the Cubo-Futurists in Russia, Natalia Goncharova frequently included wheels, gears, and other parts of machinery in her work by 1913, such as *Weaver* (Cardiff, National Museum of Wales) and *Airplane over Train* (Kazan, Art Museum). Olga Rozanova similarly employed elements of machinery, as in her *Metronome* of 1913–14 (Moscow, State Tretyakov Gallery). The association of machinery with modernism didn't require the latest kind of industrial form. Goncharova's mechanized loom and Rosanova's metronome had been around for a century, and Kasimir Malevich's *Knife Grinder* (1912, Yale University Art Gallery) uses a very old machine to symbolize his modernist "principle of glittering" (written on the back of the canvas). Russian artists also made numerous mechanomorphic forms, including Malevich's costumes for *Victory over the Sun* (1913) and the Paris-based Alexander Archipenko's *Médrano* (1915, New York, Guggenheim Museum). Their forms, by the way, have an engaging sense of humor in contrast to the rather horrifying mechanomorphic inventions of the British Vorticists and the Italian Futurists.

War and Revolution

With the onset of World War I in August, 1914, the machine took on a new and terrifying presence, so much so that the war itself (and postwar references to it) was embodied in the airplane, aerial balloons, the submarine, and motorized vehicles of all sorts, as well as in the ubiquitous cannons, machine guns, barbed wire, and gas masks. The design of airplanes and airborne armaments evolved with great rapidity, as did that of both military and civilian motor vehicles. Military needs also greatly increased the use and awareness of radio transmissions and the telephone. Before the war ended, illustrated journals were full of praise for the new designs, rightly associated with the needful economies of wartime industry. In this perspective, old-fashioned handcraft seemed anachronistic in both artistic and social terms. It was now far easier to equate modernity with industrial forms than before the war.

In Russia the war meant a violent break with the Tsarist past, and some artists claimed a stake in the future by getting rid of all representations of the real world, after passing through several years of Cubo-Futurism during which they had retained representational clues. Naturalistic representation was associated with the Tsarist past, whereas the new language of abstract art, established by 1915, could speak for the future. Kasimir Malevich's *Suprematist Composition, Airplane Flying* (figure 2), exhibited in December, 1915, is a case in point. In Delaunay's painting (figure 1) the images of airplanes are still needed, but in Malevich's, the dark rectangles at the bottom of the frame suggest the weight of earthbound forms, the yellow ones float above these, and our eye is led upward to the far-away red stripe. When writing about his art, Malevich referred to airplanes and to extraterrestrial satellites. "We are the highest point in the race of contemporary life, the kingdom of machines and motors and their work on earth and in space" (Malevich, 1918, p. 63). He likened modern technology's defeat of earth-born gravity to his self-invented Suprematist forms, which he declared to be a new visual language utterly free of references to the natural world. Similarly,

FIGURE 2. Kasimir Malevich, *Suprematist Composition: Airplane Flying*, 1915 [dated 1914]. Oil, 58.1 x 48.3 cm. New York, Museum of Modern Art.

Vladimir Tatlin's sculptures shown in the same 1915 exhibition were entirely abstract, and one *Corner Relief* (subsequently lost) was suspended on wires to free it metaphorically from contact with the earth.

The Russian Revolution of 1917, born of wartime events, led to the adoption by some party leaders of the radical language of abstraction (see Fitzpatrick, 1971; Gray, 1986; Lodder, 1983;

Henry Art Gallery, 1990, and Guggenheim Museum, 1992). Aware of the value of cultivating the artistic avant-garde, they reached out to wielders of a new visual language. In 1918, for the first anniversary of the October Revolution, Nathan Altman designed the speakers' rostrum in Saint Petersburg by surrounding the old column of Alexander with gigantic geometric forms, a new abstract language for a new society. Two years later, Vladimir Tatlin unveiled his towering *Monument to the Third International* (known through photographs and drawings) in the lobby of the meeting place of world Communism. It was a model for a central government structure to be built in glass and steel, those products of modern industry that take over the role of stone, brick, and wood. Inside the external spiral, which symbolized the ever-evolving social forms of revolution, there were to be three rooms that would revolve on electric motors. A huge cube, which would rotate once a year, housed the legislature. Above it a pyramid, housing the executive, would rotate once a month, while at the top a cylinder, enclosing the agitation and information center, would revolve once a day. The Revolution was to be served not by traditional Euclidean forms but by visionary geometry that would transform the world thanks to modern engineering. A second version of "The Tower," as it's called, was the centerpiece of the USSR's exhibition in the world's fair of 1925 in Paris: an official sanction of its symbolic language.

For Soviet revolutionary artists (we now group them together as Constructivists, although in the polemics of the time this term was a more limited one), the word *art* smacked of products isolated from the masses by faulty bourgeois conceptions. *Art* designated objects for private pleasure, so radical artists instead talked about "laboratory experiments" (Lodder, 1983). Artists were now frequently called "artist-engineers," and their work was to be closely related to factory production. The critic Boris Arvatov wrote that the progressive artist "is not merely a performer but a constructor-inventor; he can give engineering a higher creative form" (Arvatov, 1923, p. 47). Engineering, that is, cannot stand by itself

FIGURE 3. Kasimir Medunetsky, *Spatial Construction*, 1919. Tin, brass, steel and painted iron, 46 cm. high. Yale University Art Gallery, Collection Société Anonyme.

but needs this new specialist of form, a "constructor-inventor." Thanks to his mastery of visual form, the constructor supplies that which the ordinary technician lacks.

Kasimir Medunetsky's *Spatial Composition* (figure 3), exhibited with the Society of Young Artists in 1921, shows the artist-engineer at work. His deceptively simple shapes, in four different metals, form a subtle set of relationships. From one corner of the cubical base of painted metal, a brass triangle slants upward. It passes through a steel ring to reach an *S*-shaped strip of tin. To the upper end of the tin strip is attached an iron parabola, suitably painted red to inhibit rust. This parabola curves up and then down, piercing the brass triangle and reaching the other end of the *S*-strip. It forms a spatial plane that controls all the forms, each of which touches the base at only a tiny point.

Because Medunetsky borrowed his materials directly from industry, leaving them without any marks of his own touch, his sculpture seems worlds apart from Raymond Duchamp-Villon's *Horse* of 1914, the major sculpture of Epic Cubism that in its day seemed so radical. The difference between them—only seven years separate them—is one measure of the extraordinary gap that the war had opened out in modern culture. Duchamp-Villon not only intended to use bronze, the traditional sculptor's material (his plaster was cast posthumously after his wartime death), he also merged images of machine and horse. On both sides of his sculpture rise simulacra of crankshafts, and in the back is a ring of plugs that resemble the chuck of a lathe. For Medunetsky, following war and revolution, this would have been an old-fashioned object that joined machine and organism into a "romantic" piece of art. It provided no lessons to engineers on the logically organized yet inspirational uses of their materials.

Many Russian Constructivists treated their function as an equivalent to that of Communist party leaders. They would direct workers and peasants away from handcrafts towards the acceptance of machine production. Both artists and party cadres sought new socially efficient forms, forms that could embody modern industrial materials and techniques rather than hated individualism and naturalism, associated with the defunct regime of the Tsars. Artists and party leaders alike had the same view of World War I: it destroyed the old—in art as in society—to make way for the new.

Art in Russia wasn't as isolated as its rhetoric sometimes suggested. In 1919 Walter Gropius had become the director of the Bauhaus, reorienting Weimar's earlier design school in a moment of postwar socialist euphoria. By 1923 that euphoria had waned, and the school was transforming its utopian mandate to one that would wed design to enlightened capitalist commerce and industry. The Bauhaus emphasized not "art," but the examination of the possibilities inherent in industrial materials, sometimes echoing the Russian "laboratory experiments." It's true that Wassily Kandinsky and Paul Klee had studios at the Bauhaus and continued their individual fine arts work, but their teaching was subordinated to the school's investigation of new principles of design in all media. Bauhaus exhibitions were devoted to products and materials of industrial and commercial utility, and by the mid-1920s some designs entered production, ranging from Marcel Breuer's "Wassily" chair to Marianne Brandt's Kandem bedlamp (see Wingler, 1969; Humblet, 1980, and Forgács, 1995).

The French "Rappel à l'Ordre"

In postwar France there was no equivalent to the Bauhaus or the Constructivists' schooling. New energies came from diverse vanguard artists, but victorious France underwent no upheaval comparable to those in Germany and Russia; its government stressed continuity, not radical change. For the overt endorsement of modern technology, we should turn to Fernand Léger and to his allies Le Corbusier and Amadée Ozenfant who founded Purism in 1918. Purism's importance was enhanced by the influential review *L'Esprit nouveau,* which Ozenfant and Le Corbusier launched in 1920. Lasting until 1925, it was one of the major outlets for the *"Rappel à l'Ordre"* (Call to Order) that characterized the postwar years in the Parisian avant-garde. ("Rappel" carries the meaning of recall, or return, hence "Retour à l'Ordre" [Return to Order] was an alternative phrase.)

Kenneth Silver (1977 and 1989) has shown the close connection of wartime and postwar rhetoric in France to that of the Call to Order, a vanguard embrace of modern industry with aesthetic clarity that had long been attached to the classicizing underpinnings of French culture, as contrasted to supposed Germanic emotionalism and individualistic expressiveness. Le Corbusier, for example, like Malevich, used the airplane to attach wartime inventiveness to modernity. The airplane, he wrote in *Vers une architecture* (1923), "mobilized invention, intelligence, and daring: imagination and cold reason. It's the same spirit that built the Parthenon." He also explicitly linked Roman engineering with the ethos of French reconstruction, proposing that the word *Roman* meant "Unity of operation, a clear aim in view, classification of the various parts" (both citations from Silver, 1977, p. 56).

Like the Bauhaus artists and Constructivists, the Purists and Léger derived a lot of their energies from World War I but, lacking the outlets provided by a radicalized society, they poured their reactions into the broad stream of postwar reconstruction. They occupied the familiar role of avant-garde artists whose work in a stable society is provocative, perhaps leading to reform, but isn't revolutionary. Léger's friend and sometime collaborator Le Corbusier, as Mary McLeod demonstrates (McLeod, 1983), was an enthusiast for Taylorism, the American concept of rationalized

serial production that he deemed essential for France's vast program of reconstruction. Taylorism was also put forward in Soviet
Russia (Fitzpatrick, 1971) but there it could be folded into party-
managed industry, whereas Le Corbusier could only imagine it in
the context of progressive capitalism.

At the center of the Call to Order, Léger deserves special attention (see Léger, 1973; Green, 1976 and 1987; Silver, 1977 and
1989, and Herbert, 1980). His art of the early 1920s insisted upon
a close relationship of art and industry. His *Animated Landscape* of
1921 (figure 4) is a utopian vision for the modern world, one of a
large number bearing the same title that speak for an optimistic
reading of postwar reconstruction. He embraced continuity by
retaining representational imagery, discarded by Russian Constructivists who were so anxious to cut all ties with the past.
Triumphing over nature is a billboard city (Léger loved billboards
and disliked "do-gooders" who opposed them; Léger, 1973,
p. 12), for the countryside is dedicated to supplying its wants.

FIGURE 4. Fernand Léger, *Animated Landscape*, 1921. Oil, 65 x 92 cm. Private
collection. Photograph courtesy of the late Nathan Cummings.

Cows ruminate peacefully in the distance near buildings that suggest industrialized agriculture, organized to feed the city. City and
country share the landscape, because their premodern separation
has been bridged by the logic of industrial order. It's the same
industrial organization, the fruit of human rationality, that lets
men stand passively by the city, freed from the sweat of manual
labor by the power of mind over raw nature. Earned leisure is the
triumph of modernity, signaled here by the domestic coupling of
man and dog, a masculine counterpart to Léger's frequent treatment of equally impersonal women enjoying household leisure.

Like Constructivist and Bauhaus geometry, Léger's forms don't
embody the appearance but instead the logic of machine production. His numerous essays frequently make the point that to
imitate machines would be to indulge in mere descriptive naturalism. He is therefore rather close to the Russian Constructivists
despite his representational imagery. He wasn't then a member of
any party as far as we know but he was on the political left (if only,
in Soviet terms, a bourgeois liberal). He put ordinary artisans at
the center of his social credo. In his essay of 1924, "The Machine
Aesthetic," he wrote:

> The artisan regains his place, which he should always have kept,
> for he is the true creator. It's he who daily, modestly, uncon-
> sciously creates and invents the pretty trinkets and beautiful
> machines that enable us to live. His unconsciousness saves him.
> The vast majority of professional artists are detestable for their
> individual pride and their self-consciousness; they make every-
> thing wither (Léger, 1973, p. 59).

Like the Constructivists, therefore, Léger aligned himself with
democratic forces, opposing elite artists by taking on the role of a
vanguard leader who works on behalf of disfranchised artisans
and workers.[10]

The Machine Aesthetic

Léger, the Russian Constructivists, and Bauhaus artists (and more distantly, the American "Precisionists" Charles Demuth and Charles Sheeler) were allied by the *machine aesthetic*, a concept already in vogue by the time Léger used it in 1924. For them art should be the expression of industrial culture. Its forms and techniques were said to be analogous to machine production because both art and machinery process raw materials for new forms and uses. Machines do not automatically have economical structures or suitably modern appearance, so industry needs specialists in form who can instruct both engineers and public in the logic and beauty of rational means. Geometric form is the basis both of the new visual language and of industrial culture, because of the long-standing associations that define it in Western culture: measurement, order, rationality, construction, and impersonality. It's a concept that is shared by society, and therefore cannot be claimed as unique to any individual. Thanks to geometry, the *impersonality* of artistic forms guarantees a visual language suitable to mass production and social engineering, hence the unmarked industrial surfaces of Medunetsky's *Spatial Construction* and the impersonality of Léger's images are programmatic assertions.

For adherents of the machine aesthetic, nature and naturalistic art were equated with prewar society. In the postwar era, humans can triumph over nature thanks to the new technologies; they must not continue merely to follow nature but instead must transform her. And it *is* "her" in this very masculine set of beliefs. "The contemporary environment is clearly the manufactured and 'mechanical' object; this is slowly subjugating the breasts and curves of woman, fruit, the soft landscape—inspiration of painters since art began" (Léger, cited in Green and Golding, 1970). Postwar reconstruction made idealists into "constructors" and "artist-engineers," not mere artists.

The least common denominators of the international machine aesthetic can be summarized under six headings. Geometry is their common expression, which is why geometric abstraction had such powerful resonance in the postwar era.

Construction versus Destruction

The postwar world requires the destruction of prewar hierarchies. The war had a cleansing effect because it destroyed monarchy and naturalism in the arts. Reconstruction is based upon violent upheaval; progressive politics and progressive art are allies.

Rationality

A social term of broad currency, *rationality* enfolds the aesthetic associations of reason, order, utility, and economy. It speaks for the conscious and man-made in opposition to the subjective and private (associated with the prewar era), and for the secular, because geometric harmony is no longer conceived as a revelation of the divine. Men create, they don't "discover" or "reveal."

Triumph over Nature and Naturalism

Nature is equated with the traditional, the accidental, the irregular (divine purpose is no longer believed in), and the imitative, as distinct from secular modernism in which geometry represents industrial triumph over the "natural" forms of prewar social organization (monarchies were defeated).

Anonymity, Impersonality, Anti-individualism

The social and artistic enemy is individualism and emotive expressionism, which are antisocial. The anonymous worker and *le peuple* are to be celebrated. Because modern humans use machines as tools in social cooperation, personal brushwork in art and handmade irregularities in industry must be discarded. The anonymous in art and the impersonal in machinery, symbolized by geometric form, are the signs of the subordination of the individual to social good.

Aspiration toward Architecture

The machine aesthetic in painting and sculpture claimed architecture as the favored sister art, in contrast to the prewar vanguard's preference for music, now considered too personal. Postwar reconstruction obviously favored the most socially useful of the arts, not just in its accomplished structures, but also in its production by teams of designers and workers: the quintessential social collaboration. Léger titled one of his pictures *Architecture* (1923, private collection) and El Lissitzky's pictures and graphics, no matter how abstract, are full of three-dimensional simulacra of architectural forms.

Reproducibility, Economy of Means

Mass production must triumph over isolated handcrafts and individualism to bring the largest benefits to the most people. Geometric forms in industry and in art embody the logic of reproducibility and its social benefits. Although Léger and the Constructivists crafted their work by hand, they gave it the appearance of impersonality to proclaim the suppression of the self. The social value of their art lies in its apparent suitability to mass production. Logically, as the prescient Meyer Schapiro realized in 1937, the aesthetic terminology of abstraction was borrowed from economics and industry. After World War I, he wrote,

> the older categories of art were translated into the language of modern technology; the essential was identified with the efficient, the unit with the standardized element, texture with new materials, representation with photography, drawing with the ruled or mechanically traced line, color with the flat coat of paint and design with the model or the instructing plan (Schapiro, 1937, p. 97).

The Dada Opposition

Until now in this paper I've been discussing artists whose work embraced positive attitudes toward humankind's ability to use new industrial techniques to alter both art and society. Other modernists, however, adopted opposed viewpoints, and we must now consider them. Leading the way before 1918 was Marcel Duchamp, that heir of nineteenth-century Dandyism whose mocking attitude towards rationality and towards art itself has come to be regarded as a fundamental manifestation of modernism (see Camfield, 1987; Düve, 1991; Hultén, 1993, and Naumann, 1994). Seeking alternatives to conventional painting after 1912, he created objects that could only be understood in terms of the conceptions they led to when their appearance was linked with Duchamp's words for them. Word and concept, that is, were inextricably fused with the visual to escape what Duchamp regarded as the limitation of "retinal art."

One of his best-known objects is *Fountain,* a porcelain urinal turned on its back but unaltered except for his signature as "R. Mutt" (Camfield, 1987). In 1917 it was rejected by a vanguard exhibition society in New York (and was subsequently lost), presumably because of its potential for scandal but also because it didn't seem to be "art" at all. Duchamp's radical act (appreciated much more in the 1960s than in 1917) was to declare that art was pure concept, that an ordinary manufactured object could be transformed by that declaration, and that "originality" didn't require artistic handwork. Two years earlier Duchamp had begun working on his large *Bride Stripped Bare by Her Bachelors, Even* (Philadelphia Museum of Art), a work on glass "uncompleted" in 1923, and then repaired in 1936 to incorporate prominent cracks in its glass. Readable only if one is privy to some of the many explanatory notes Duchamp published, the "Large Glass" is a structure of frustrated mechanical sex, a complicated and immensely rich mechanomorphic world that became an object of pilgrimage in the 1960s.

Duchamp's work was at the center of what came to be known as "New York Dada," notable for its frequent assaults upon conventional ideas about the machine world and about art (Naumann, 1994). Morton Schamberg exhibited metal pipes in a miter box as *God* (1918, Philadelphia Museum of Art), and Duchamp's French sidekick Francis Picabia made paintings and drawings using the imagery and style of diagrams of machines. Some were "portraits" of notable figures (Alfred Stieglitz was rendered as a diagrammatic camera), others were instances of mechanical sex rendered in gears and pistons, including Picabia's *Amorous Parade* (1917, Chicago, Mr. and Mrs. Morton G. Neumann).

Although Duchamp's prewar work had adumbrated the Dada spirit, it was only during World War I that Dada became established as a broad if loosely coordinated activity. The name itself was provided in 1916 by the group led by Hugo Ball at the Café Voltaire in Zürich (Hans Arp, Sophie Tauber, Tristan Tzara), and adopted the next year by Duchamp, Picabia, and their American colleagues (see Richter, 1965; Rubin, 1969; Ades, 1978; Lewis, 1988; Dachy, 1990, and Naumann, 1994). The New York and Zürich groups retreated from the war but instead of joining anti-war groups they adopted attitudes of sardonic withdrawal, acting out the belief that in face of war's monstrosities, the only sanity was unreason.

In the aftermath of the war other Dadaists treated the machine as the agent of destructive violence, and insisted that rationality was inherently oppressive. Various manifestations of international Dada rapidly spread from New York and Zürich to major centers of European culture. It became a notorious phenomenon of postwar culture until about 1924, when Surrealism took over and transformed Dada's waning energies. Dada wasn't an organized movement, but its disparate practitioners had similar sardonic, often bitter reactions to the consequences of mechanization, particularly in light of the war.

It was in postwar Germany that Dada gained the largest number of adherents, doubtless because there the mood of defeat and upheaval gave greater outlet to antiwar and antirational beliefs than was true in victorious France and Communist Russia. Already in 1918 a group of Dadaists in Berlin had linked Dada with the war. Listen to the echoes of war and postwar trauma in the language of their manifesto:

> Life appears as a simultaneous muddle of noises, colors and spiritual rhythms, which is taken unmodified, with all the sensational screams and fevers of its reckless everyday psyche and with all its brutal reality. . . . Dada is the international expression of our times, the great rebellion of artistic movements, the artistic reflex of all these offensives, peace congresses, riots (cited in Richter, 1965, p. 106).

In the first major Dada exhibition in Berlin, in 1920, a pig in military uniform dangled from the ceiling, near a large painting by Otto Dix of *War Cripples* (formerly Dresden, Stadtmuseum), a savage indictment of war that made a mockery of any hopes for postwar recovery. Elsewhere in that exhibition was a huge collage by Hannah Höch, *Cut with the kitchen knife Dada through the last Weimar beer belly cultural epoch of Germany* (Berlin, National Gallery)(see Lavin, 1993). Höch's cutouts from the mass press echo "these offensives, peace congresses, riots," and the overall rhythms of her polemical collage express "screams and fevers." In the upper left, there's the head of Einstein from which spring machine forms and the word *Dada.* In the upper right is the deposed Kaiser Wilhelm surrounded by Weimar political and military leaders; his moustache consists of two wrestlers. Machine forms abound throughout the collage, countered by many images of women, shown as the antithesis of masculine machinery and military culture (it's a female who "cuts with the kitchen knife"). As Maud Lavin has shown (1993), negative and positive connotations swarm throughout the collage: images of both human and machine movement, excitement, and riotous liberation, not just "antimachine" connotations.

Photo-collage was a subversive handmade form particularly prominent among German Dadaists, notably Höch, John Heartfield, and Raoul Haussman. In contrast to Picasso, who avoided photographic reproductions in his prewar collages, and who personalized his borrowings with his own drawing, the German Dadaists rarely added any drawing of their own to their montages, instead using clippings from the machine-based mass media to promote resistance and activism. In doing so, they attacked the old concept of "originality" and the wholeness of earlier art, and therefore assaulted the supposedly rational means of representation itself, not just the rationality of industry and mechanized war. (It's worthwhile remembering that Léger held the opposite point of view, and therefore never used photographic reproduction nor, indeed, collage.)

Max Ernst

Max Ernst, less overtly political than the Berlin Dadaists, introduced a technique closely related to photo-collage and equally subversive. From 1919 through 1921 he made a number of Dada works by taking sheets from a compendious publication of 1914 whose engravings represent all sorts of objects, charts, and images used in school instruction.[11] He suppressed some of their objects and altered others by coloring the sheets with gouache, paint and drawing, occasionally pasting on fragments borrowed from this and other publications. For *Hydrometric Demonstration of Killing by Temperature* (1920, private collection), Ernst converted a page of laboratory equipment into a compartment of machinelike objects that protrude into a blue sky. This bizarre composition looks like a premonition of Hitler's crematoria, but in 1920 it would have conjured up a menacing factory. *Two Ambiguous Figures* (1919, private collection) transforms another sheet into mechanomorphic creatures sporting goggles reminiscent of gas masks.

Showing greater distance from wartime invocations, *Sheep* of 1921 converts a group of geometric objects used in primary education to a strange vision (figures 5 and 6). Set in a desert landscape where a reindeer sled disappears in the distance, these objects originally dedicated to rational investigation (object

FIGURE 5. Max Ernst, *Sheep*, 1921. Gouache and collage over catalogue page, 11.2 x 16 cm., for Paul Eluard, *Répétitions* (Paris, 1922). Paris, Musée national d'art moderne.

lessons!) have become otherworldly forms watched over by an ancient warrior. Of the two flanking arms, one mysteriously brandishes a tube and petcock, while the other is flayed, its horror partly mitigated by recognition that it was borrowed from a medical chart. Three years before Surrealism was founded, we are already in a world of unreason, all the more upsetting because the very instruments of rational discourse are used to deny the logic of construction and understanding.

In paintings unaided by collage, and therefore more like the procedures of traditional art, Ernst pursued the same attack upon reason and, by extension, upon the mechanical world. The central image of *Elephant Celebes* of 1921 (figure 7) was inspired by a photograph of a double-legged structure of dried clay made by the Konkomba people of western Sudan to store their grain (Penrose, 1972, pp. 14–15). Ernst's elephantine form would defy the descriptive powers of the group of blind men. The Africanness of its "primitive" shape, so much the other of modern industry, is mixed with its opposite, mechanistic forms. At the top, geometric shapes recall both the "Metaphysical Art" of Giorgio de Chirico of the prior decade and the turrets of armored vehicles of World War I.[12] The monstrous "trunk" looks like industrial tubing, but terminates in an organic image of a bull's head and napkin, perhaps threatening the woman in the lower right, or already feasting on her absent head. Paradoxically, the eyes of the bull can be read as a headless woman's breasts, and the horns as her arms. Either reading sets up a dialogue with the decapitated nude, whose

FIGURE 6. *Kölner Lehrmittel-Anstalt* (Cologne: Hugo Inderau, 1914), p. 236.

mutilated form also resonates strangely with the phallic tower behind. The age-old theme of "the world upside down" appears in the two fish who swim in the sky, but the downward spiraling smoke to the right makes us think of that modern wartime disaster, a stricken airplane or observation balloon, all the more so because the landscape resembles an airfield.

No such monster and no headless nudes could ever disturb the balanced and optimistic worlds that Léger created. He and Max Ernst are a study in unparallel lives, for although both lived in the muck and mire of World War I trenches, witnesses to untold slaughter of their comrades, they established opposed viewpoints in their art. The earth of Léger's *Animated Landscape* isn't an airfield for a clanking monster, but a productive countryside that will feed his billboard city. His robotic figures are not mutilated, but constructed of impersonal geometry that stands for the positive

FIGURE 7. Max Ernst, *Elephant Celebes*, 1921. Oil, 125 x 108 cm. London, Tate Gallery.

capacities of human reason to replace hard labor with rational instruments. He will orchestrate material forces towards a new social harmony in which Ernst's irrational inventions could have no place.

For Léger, in victorious France, and for the Russian Constructivists, in the midst of building a new political order, the dehumanization of war was a spur toward ideals of social reconstruction, as it was for Gropius and the Bauhaus. By contrast, for Ernst and many other artists in Germany, the dehumanization of wartime violence and civilian suffering was tragic proof that humans will use coercive social order, impersonality, and the machine to destroy life. The wondrous potential of modern technology therefore had to be countered with its dangers. The Dadaists could accuse Léger, the Russian Constructivists, and the Bauhaus of serving both Capitalism and Communism by making the machine into an instrument of social order. The radical nature of Dada art is to have exposed the social consequences of the machine world, that social order can be imposed to become social control. This is an essentially political stance because it recognized that the machine wasn't the classless instrument of a better world but a symbol of the control of some humans by others. The machine and other emblems of reason caricatured by the Dadaists spoke for the lack of control by the victims of social order.

The Machine Then and Now

Until now in this paper I've marked out the gap between those who greeted the arrival of the modern machine as the harbinger of a new utopia, and those who looked upon it with despair. From Leo Marx (1964) and Langdon Winner (1977), we have long been aware that this opposition has characterized modern culture since the onset of the industrial revolution. All I've hoped to do is to document that critical moment revolving around World War I when modernists elevated it to the dominant structures and subjects of painting and sculpture.

It's true that there were some crosscurrents between the Dadaists and the technological optimists of the machine aesthetic. Both sides believed that old ideas had been destroyed by the war and that traditional naturalism could not suit the new era that humankind was entering in the postwar years. Dada artists were not unrelievedly against all ramifications of the machine.

Hannah Höch and Raoul Haussman emphasized its uses for
destruction and social control but they were also struck by the
potential of modern machines to release the human imagination.
Some Dada artists drew close to Constructivist geometry, particularly Kurt Schwitters, whose collages and reliefs after 1921 often
took on a regularity bespeaking rational organization. On the
other hand, the attacks that Dada made on the old order were so
engaging that a number of artists associated with the machine aesthetic flirted with it. Before he went to the Bauhaus, Laszlo
Moholy-Nagy composed a number of frankly Dada works, and
Alexander Rodchenko's photo-montages for Mayakovsky's poem
Pro Eto (1923) have a decided Dada exuberance. However,
although these crosscurrents reflect mutual awareness among the
avant-garde, they do not deny the broad oppositions that I've
sketched out.

Although each prior generation had its machine optimists and
detractors, it was only in the turbulent second decade of our century that modernists felt obliged to deal head-on with the arrival
of the machine, particularly in the aftermath of World War I. In
fact, the visit this paper has made to art of seventy-five years ago
seems to take us back to a kind of naive or archaic world. Since
then, Surrealism, science fiction, frequent wars, apocalyptic
atomic power, and microcomputing have all intervened to prevent us from seeing into technology with such doubtful clarity.

For us today, technology's intersections with artistic culture
are so varied and intertwined that those old polarities no longer
seem a sufficient description of modern beliefs. The astonishing
rise of science fiction has been possible because of the widespread
belief that modern technology, symbolized by machines, gives
humans the capacity to build whole new worlds but also to
destroy, the two capacities in dangerous alliance, no longer readily separable. In fact, science fiction, although its origins predate
our century, only became a major element in advanced industrial
culture after World War I, its rise to prominence roughly coinciding with the development of Surrealism.

In our present culture, many of us actually share the ideas embodied in both Ernst and Léger. We place great hopes in the latest advances of technology. We travel about in its wondrous machines, but we find in current films, television, and computer games an unending display of high-tech machines dedicated to violent ends. In science fiction, we're nearly always under the threat of imminent violence, and can find images that more or less look like both Léger and Ernst. The separation of illusion from reality, of human organisms from machines, is no longer evident, as we can see when we walk by the penny arcades in our shopping malls or read about sophisticated work on cognitive computing machines.

Notes

1 For this essay I am generally indebted to Mumford, 1934; Schapiro, 1937; Francastel, 1956; Marx, 1964; Winner, 1977; and Henderson, 1983. For the era just before and after World War I, I've learned especially from Herf, 1984; Silver, 1977 and 1989; and Cork, 1994. I look forward to published papers from the enterprising symposium *From Energy to Information: Representation in Science, Art, and Literature,* April 3–5, 1997, at the University of Texas, Austin, organized by Linda Dalrymple Henderson, Bruce Clarke, and Richard Shiff.

2 Irony of ironies: a branch railway bisected Monet's property; he built a tunnel under it to reach his water-garden. Few historians mention the railway lest the juxtaposition interfere with the nature poetry of the waterlilies.

3 Veblen 1899, p. 114 and passim. I have pointed out the appropriateness of Veblen in this regard in "Impressionism, Originality, and Laissez-faire," *Radical History Review* 38 (1987): 7–15.

4 It is Daniel Robbins who defined Epic Cubism as the work of Gleizes, Delaunay, Le Fauconnier, and others whose concern for programmatic themes distinguishes them from Braque and Picasso. He coined the term in his exhibition catalogue, *Albert Gleizes 1881–1955* (New York: Guggenheim Museum, 1964), and pursued the concept subsequently in publications on Jacques Villon, Jean Metzinger, and Henri Le Fauconnier.

5 See the entry on Gleizes by the late Daniel Robbins in Herbert, Apter, and Kenney, eds., 1984, pp. 301–303.

6 *Le petit atelier de mécanique*, 1913 (Washington, Phillips Collection); *L'atelier de mécanique*, 1914 (New York, private collection); *L'atelier de mécanique*, 1914 (Columbus, Ohio, Columbus Gallery of Fine Arts).

7 Delaunay's writings defining "simultaneity" and other issues were edited by Pierre Francastel, Robert Delaunay, *Du Cubisme à l'art abstrait, documents inédits* (Paris: 1957). For the Bergsonian import of these ideas, see Antliff, 1993, pp. 40–41, 50, 53–54, 108, and 132. We should note that Delaunay's target-like forms also recall the multicolored disks figured in books on color theory (one of Delaunay's preoccupations). Spun at high speed by simple machines, the disks' colors merged, hence permitting a quantification of the constituent hues.

8 The phrase linking a car with the Greek sculpture is usually cited without including the phrase "that seems to ride on grapeshot," thereby softening Marinetti's warlike imagery. William Valerio's unpublished dissertation ("Boccioni's Fist: Italian Futurism and the Construction of Fascist Modernism" [Yale University, 1996]) is a brilliant analysis of the relationship of violence and protofascist politics with the aggressive masculinity and partly repressed homoeroticism of the Futurists.

9 Because Epstein stripped the sculpture of its metal drill when he made the later bronze in 1916, Richard Cork (1994, pp. 134–35) interprets it as the supplanting of prewar optimism by a sense of impotence and forlornness in face of the war's disasters.

10 By the standards of Marxism, Léger's definitions of artisan and worker are naive. It's true that unlike Picasso and most other contemporary vanguardists he often painted men identified as factory workers, but in his writings he included among artisans the modest shopkeeper who designed his own window displays, and he repeatedly endorsed capitalist competition as the essential structure of society (Léger, 1973, essays of the 1920s).

11 This was the sales catalogue *Kölner Lehrmittel-Anstalt* published by the Hugo Inderau firm, an exhibitor in the industrial exhibition of 1914 in Cologne. Ernst's use of the catalogue (he made twenty-five overpaintings from it) is documented by Dirk Teuber in Herzogenrath, 1980, pp. 206–274. For Ernst, see also Penrose, 1972; Spies, 1991; and Camfield, 1993.

12 I am indebted to Richard Cork (Cork, 1994, pp. 258–60) in my observations of *Elephant Celebes*, although I think he pushes analogies with the war rather too far.

References

Ades, Dawn, *Dada and Surrealism Revisited* (London: Arts Council of Great Britain, 1978).

Antliff, Mark, *Inventing Bergson* (Princeton University Press, 1993).

Arvatov, Boris, *Art and Class* (1923), excerpt in *The Tradition of Constructivism,* Stephen Bann, ed. (New York: Viking Press, 1974), p. 47.

Camfield, William, *Marcel Duchamp Fountain* (Houston: Menil Collection, 1987).

Camfield, William, *Max Ernst: Dada and the Dawn of Surrealism* (Munich and New York: Prestel, 1993).

Cork, Richard, *Vorticism and Abstract Art in the First Machine Age,* 2 vols. (Berkeley: University of California Press, 1976).

Cork, Richard, *A Bitter Truth: Avant-Garde Art and the Great War* (New Haven and London: Yale University Press, 1994).

Dachy, Marc, *The Dada Movement, 1915–1923* (New York: Rizzoli, 1990).

Delaunay, Robert, *Robert Delaunay, Du Cubisme à l'art abstrait, documents inédits,* Pierre Francastel, ed. (Paris: 1957).

Düve, Thierry de, ed., *The Definitely Unfinished Marcel Duchamp* (Cambridge: MIT Press, 1991).

Fitzpatrick, Sheila, *The Commissariat of Enlightenment: Soviet Organisation of Education and the Arts underr Lunacharsky, October 1917–1921* (Cambridge and New York: Cambridge University Press, 1971).

Forgács, Eva, *The Bauhaus Idea and Bauhaus Politics,* John Bátki, trans. (Budapest: Central European University Press, 1995).

Francastel, Pierre, *Art et technique aux XIXe et XXe siècles* (Paris: Editions de Minuit, 1956).

Gray, Camilla, *The Russian Experiment in Art 1863–1922,* Marian Burleigh-Motley, ed. (London: Thames and Hudson, 1986).

Green, Christopher, and Golding, John, *Léger and Purist Paris* (London: Tate Gallery, 1970).

Green, Christopher, *Léger and the Avant-Garde* (New Haven and London: Yale University Press, 1976).

Green, Christopher, *Cubism and Its Enemies: Modern Movements and Reaction in French Art 1916–1928* (New Haven and London: Yale University Press, 1987).

Guggenheim Museum, *The Great Utopia: The Russian and Soviet Avant-Garde, 1915–1932* (New York: Rizzoli, 1992).

Henderson, Linda, *The Fourth Dimension and Non-Euclidean Geometry in Modern Art* (Princeton: Princeton University Press, 1983).

Henry Art Gallery, *Art into Life: Russian Constructivism 1914–1932* (Seattle: University of Washington Press, 1990).

Herbert, Robert L., *Léger's Le Grand Déjeuner* (Minneapolis Institute of Arts, 1980).

Herbert, Robert L., Eleanor S. Apter, and Elise K. Kenney, eds., *The Société Anonyme and the Dreier Bequest at Yale University* (New Haven and London: Yale University Press, 1984).

Herf, Jeffrey, *Reactionary Modernism: Technology, Culture, and Politics in Weimar and the Third Reich* (Cambridge and New York: Cambridge University Press, 1984).

Herzogenrath, Wulf, ed., *Max Ernst in Köln, Die rheinische Kunstszene bis 1922* (Cologne: Kölnischer Kunstverein, 1980).

Hultén, Pontus, Jennifer Gough-Cooper, and Jacques Caumont, *Marcel Duchamp: Work and Life* (Vienna: Palazzo Grassi, 1993).

Humblet, Claudine, *Le Bauhaus* (Lausanne: Editions l'Age d'homme, 1980).

Hüter, Karl-Heinz, *Das Bauhaus in Weimer, Studie zur Geschichte einer deutschen Kunstschule* (Berlin: Akademie Verlag, 1976).

Lavin, Maud, *Cut with the Kitchen Knife: The Weimar Photomontages of Hannach Höch* (New Haven and London: Yale University Press, 1993).

Léger, Fernand, *Functions of Painting*, Edward F. Fry, ed., Alexandra Anderson, trans. (New York: Viking Press, 1973).

Lewis, Helena, *Dada Turns Red, the Politics of Surrealism* (Edinburgh: Edinburgh University Press, 1988).

Malevich, Kasimir, "Architecture as a Slap in the Face to Ferro-Concrete" (April 1918), in his *Essays in Art 1915–1933*, Troels Andersen, ed., Senia Glowacki-Prus, trans. (New York: George Wittenborn, 1971), p. 63.

Marinetti, Felippo, "First Futurist Manifesto" (1909), in *Futurist Manifestos*, Umbro Apollonio, ed. (New York: Viking Press, 1973), pp. 21–22.

Marx, Leo, *The Machine in the Garden* (New York: Oxford University Press, 1964).

McLeod, Mary, "'Architecture or Revolution': Taylorism, Technocracy, and Social Change," *Art Journal* 43 (Summer 1983): 132–47.

Naumann, Francis, *New York Dada, 1915–23* (New York: Abrams, 1994).

Penrose, Roland, *Max Ernst's Celebes, The 52nd Charlton Lecture* (Newcastle: University of Newcastle upon Tyne, 1972).

Richter, Hans, *Dada: Art and Anti-Art* (London: Thames and Hudson, 1965).

Rubin, William, *Dada and Surrealist Art* (London: Thames and Hudson, 1969).

Schapiro, Meyer, "Nature of Abstract Art," *Marxist Quarterly* 1 (January–March 1937): 77–98.

Silver, Kenneth, "Purism: Straightening Up After the Great War," *Artforum* 15:7 (March, 1977): 56–63.

Silver, Kenneth, *Esprit de Corps, The Art of the Parisian Avant-Garde and the First World War,* 1914–1925 (Princeton University Press, 1989).

Spies, Werner, *Max Ernst Collages: The Invention of the Surrealist Universe,* John W. Gabriel, trans. (New York: Abrams, 1991).

Teuber, Dirk, "Max Ernst's Lehrmittel," in *Max Ernst in Köln,* Wulf Herzogenrath, ed. (Cologne: Kölnischer Kunstverein, 1980), pp. 206–21.

Veblen, Thorstein, *The Theory of the Leisure Class* (New York: Macmillan, 1899).

Wingler, Hans, *The Bauhaus,* Wolfgang Jabs and Basil Gilbert, trans. (Cambridge: MIT Press, 1969).

Winner, Langdon, *Autonomous Technolgoy. Technics-out-of-Control as a Theme in Political Thought* (Cambridge: MIT Press, 1977).

PART 7

Contemporary Moral and Political Issues: A Discussion

A discussion between conference participants and the audience of some of the political and moral issues raised by the new technology among which are questions of intellectual property, democratic representation, censorship, and unequal access.

Introduction

BY ROBERT McC. ADAMS

T HREADING their way through much of our current discussion
have been two interrelated assumptions. One is of the immense,
interrelated variety of information-based, technology-enabled
changes going on all around us. The other, although somewhat
less explicit, is of the simultaneous, seemingly irresistible additiv-
ity of their effects. The first is neither surprising nor problematic;
it probably was uppermost in most discussants' minds upon coming to this conference. But is the other really a necessary corollary
of the first?

We have met in a world-metropolis where to stand still has
always been unimaginable. Most of us come out of research settings that value the discovery of new, dynamic qualities more
highly than the reaffirmation of older, more static ones. Being
matters less than becoming, it is often said in these circumstances.
Even in human affairs more generally, hope-tinged perception
outweighs accurate, dispassionate measurement as a spring for
action.

Unsurprisingly, therefore, recent discussions have largely
avoided contingencies and complexities—those already encountered and those still to be discovered—and dealt instead with
more reassuring and unproblematic visions of what is to come. Yet
in our daily lives, we cannot escape knowing that the larger condition of the world is one of appalling irrationalities, injustices,
disjunctures, and unintended consequences. In this concluding
section, therefore, I hope we can somehow bring closer together
the smooth, implicitly deterministic curves of progressive change
that are so confidently projected and the obstacles and distortions
to them that will surely arise from the convulsive, never truly predictable messiness of real clashes of interest and chains of events.

As it has been too largely in our considerations here, the dominance of the United States in what is currently under way is beyond question. Redundant confirmation could be provided if needed by all such conceivable indices as numbers of personal computers and Internet connections in use; openness and friendliness of capital markets to new, high-technology start-ups; density of active networks and circuitry; or the prevailing directionality of information flows. But how long will this dominance remain unchallenged? To what extent does it rest on a cycle of unevenly distributed prosperity and growth in productivity that rising market demand alone cannot indefinitely sustain?

Matching the uncertainties of technological outcomes are others affecting corporate concentrations of power. Especially in intensely competitive, high-technology fields where global research and development, production, and marketing strategies are essential, giant firms have won substantial freedom from national constraints. Similarly unrestrained concentrations of corporate control now are reaching deeply into the media, affecting sports events, entertainment, and even what passes these days for news. But how far is the international homogenization of culture in our American image really likely to continue before it generates forces in other countries strong enough to arrest the trend?

This is a likely arena of further conflict and surprising outcomes that demands further study. Only in science, in my judgment, is the case for a new, truly global, information-based system of intercommunication and interaction compelling. There alone is the long-term outcome of present trends likely to be gratifyingly close to current, linear projections of them.

To allow myself one hoped-for perception, a U.S. particularity helping to account for our present leadership is the protection afforded by our Constitution and Bill of Rights, not only for the freedom of action and association of individuals, but of private concerns construed as individuals. Whatever else the future may hold, this should prevail on a widening international scale as a powerful force for the actuation of an Information Age that rapidly and effectively balances its social, intellectual, and economic promise with a concern for distributive justice. It is a pleasure to note that our concluding section directly addresses this issue.

Technology and Democracy

BY JERRY BERMAN AND
DANIEL J. WEITZNER

> The Internet is a far more speech-enhanc-
> ing medium than print, the village green, or
> the mails. . . . (T)he Internet may fairly be
> regarded as a never ending worldwide con-
> versation.—*ALA v. Department of Justice*, 929
> F.Supp. 824 (E.D.Pa. 1996) (opinion of J.
> Dalzell)

AMERICANS have a long-established reputation of being optimistic (often naively so) about new technology, and about representative democracy. The experience of the Internet in America
offers substantial reasons to be optimistic about the positive
impact of new interactive digital media on our culture and political life. This experience compels those who care about democracy to support the continued openness of, decentralized
architecture of, and affordable access to the Internet.

The Internet has shown that both choice of architecture and
affordable cost of service are essential to enhancing citizen participation in the democratic process. To be clear, the Internet will
neither assure individual participation in democracy, nor will it
assure that the final outcome of democratic deliberations is fair
and just. The Internet is simply a platform. However, the unique
nature of this platform offers a singular opportunity to support
the renewal of citizen democracy. All who would promote a
richer, more vibrant, and more participatory democratic life
ought to pay special attention to this new medium.

The strength of the Internet derives from both affordable
access and a uniquely open architecture. Traditional communications media, such as radio and television, have been affordable
and readily available around the country, but have failed to
enable full democratic participation because of architectural lim-

371

itations. For example, on-line discussions of political issues enable users to exchange views, and even pose questions to political figures, in a way that broadcast television can never support. The Internet's architecture allows for a diversity of views and exchange of information that are simply impossible in any other communications medium.

The one certainty in the otherwise uncertain world of the Internet is that the Internet is constantly changing. The Internet of several years ago without the World Wide Web is vastly different than that Internet with the Web. The Internet changes, in large part, in response to what its users need. Preserving the open, democratic character of the Internet requires maintaining the architectural characteristics discussed here, but it also requires actually using the Internet for democratic purposes. As the Internet changes, which it is certain to do, this is the best way to assure that it changes in directions that support democratic activity.

The Internet supports such a great diversity of opinion, ideas, and information because it is a decentralized network. A user can create a new Web site or participate in a Usenet newsgroup without obtaining permission in advance from any central authority. For example, to create a Web page that will be publicly accessible to millions of Internet users around the world, one need only find an Internet-connected computer and, often, pay that operator of that computer for Web-site hosting service. The decentralized architecture of the Internet has guaranteed that there will be numerous Web-site hosts from which to chose. Unlike traditional broadcast media, the Internet has no central control point. Anyone with content to publish or ideas to exchange can do so from any point on the network.

Equally important, the resources needed to establish a new Web page or post a new idea are essentially unlimited. Adding a new Web page or newsgroup posting does not require that another site or page be eliminated from the Internet. In fact, the marginal cost of the most recent web page added to the net is equal to or less than the cost of the first page. This is an environment characterized by an abundance of communications opportunities. This provides, in turn, an abundance of opportunities for democratic participation.

This abundance stands in sharp contrast to the scarcity of channels and spectrum that has been such a prominent feature, for example, of the broadcast and cable television media. Broadcasting a program on today's television systems requires that one compete for, and usually pay a high price for, a channel slot under the control of the broadcaster or cable operator. The high demand for channels in traditional media has raised their cost far beyond the means of most community organizations and all but the wealthiest individuals. This highly constricted access to the mass media has all but strangled democratic discourse in the United States.

In sharp contrast, the Internet has opened up new opportunities for grass-roots discourse. The ease of participating in on-line discussions and even in establishing web sites or mailing lists as support for coalition activities has enable individuals and organizations to form new coalitions rapidly as issues arise. For example, over thirty organizations and one hundred thousand individuals from all around the United States formed an on-line coalition in response to the congressional effort to censor speech on the Internet. This coalition was organized and became a vocal force within days. Such an effort would have been simply impossible without the Internet.

All Internet users are able to be both speakers and listeners, publishers and readers, content providers and content consumers. The bidirectional, interactive nature of the Internet is another key attribute that makes it such a unique and effective forum for democratic discourse. Indeed, only on the proverbial town square is there a greater degree of interactive, back-and-forth communications than what is possible on-line. As advanced telecommunications access services develop, it will be essential to assure continued, upstream and downstream, interactive communications paths.

The bidirectional nature of the Internet is equally as important for democratic activity as is the decentralized architecture. Interactive capability enables Internet users to express their opinions back to the legislative process, rather than just hearing one-way communication from politicians and advocates. As an example, many "netizens" have used the Internet to create on-line petitions and letter-writing campaigns to send their opinions back to elected representatives.

From the beginning, the Internet was designed to support multiple access points. Initially, this was to meet military planners' need for a disaster-proof network. Today, we all reap the benefit of this decentralized architecture, which allows more service providers to connect to the Internet every day and make new services available. In most areas of the country, Internet access is available from a variety of sources, including small and large Internet service providers, commercial on-line services, schools and libraries, freenets and other community networks, as well as traditional bulletin board systems (BBSs) linked to the Internet.

The growth, technical advances, and increasingly wide reach of the Internet has been spurred in recent years by the vibrant, competitive market for Internet access services. Citizens and organizations that rely on the Internet have been the beneficiaries of this competitive environment. Service providers compete to offer better prices, more reliable basic access, and innovative new services such as Web-site hosting with the latest Web-server features.

The benefits of an open-access network go beyond mere price competition. Open interconnection features of the Internet assure that services will develop to meet the varying needs of the diverse Internet user community, from large corporations to small libraries, individuals, and small community organizations. A great diversity of users is made possible because a variety of service providers are able to coexist on the Internet. This breadth of users creates the potential for a true diversity of opinion and ideas in on-line forums.

An open-standards environment allows the Internet to evolve innovative new services to meet changing user needs. From year to year, the face of the Internet changes. Today, many people think of and experience the Internet as primarily the World Wide Web. Yet, two or three years ago, the Web was little more than an experimental service being developed in a physics laboratory in Switzerland. This year's World Wide Web is vastly different, and more powerful, flexible, and easy to use than it was last year. At a more technical level, Internet routing and addressing protocols are evolving in order to meet increasing usage levels around the world.

These and other technical advances have been possible because of the open, public standards on which the Internet is built. In this open environment, new technologies can be developed and deployed by various members of the Internet community. New standards that gain popularity are adopted by the Internet as a whole, while others are not. This development process, however, can proceed relatively easily because the basic Internet standards are public and available as building blocks for new developments.

The Internet's architecture is what makes it unique, but it would be of no use if access were too expensive for users. There are good reasons to believe that information on the Internet will be inherently more affordable than other media. However, as multimedia and other high-bandwidth applications become more popular the ability of the voice-call-oriented telecommunications infrastructure to maintain affordable access to the Internet will be called into question.

Three separate cost elements ought to be distinguished in this discussion: 1) the cost of a personal computer or other access hardware; 2) the cost of Internet or on-line services; and 3) the cost of the underlying telecommunications service that connects the user to the Internet or on-line service. Hardware costs remain substantial, but new developments in the market, such as WebTV, network computers, and consumer-oriented personal computers, suggest that this barrier may be easing. Internet access service costs continue to decrease in the face of a competitive market. Finally, although the low cost of basic telephone service (especially flat rate local calling) has been viewed as a significant boon to the Internet, most users around the country have no viable option other than slow, analog phone lines. Internet users need more choices—particularly architectural choices—in the market for the basic telecommunications components of Internet access in order to maintain affordable, higher-bandwidth methods of access to the network.

Full participation in on-line political life for individual citizens and community political organizations is not likely to be achieved simply through market forces. As has been the case in all other communications media, redirection of resources will be necessary to ensure that everyone is able to afford participation in the on-line commons. The structure for this support is likely, however, to be different than for other media. For example, political access to television and radio spectrum is often structured as a set-aside of channel space for the use of political candidates or parties. This approach has been necessary because of the scarcity of channels. Assuring access to the Web does not require that a portion of all Web sites be set aside for certain users. Rather, we must instead be sure that the users we seek to support have sufficient financial resources to purchase access and Web-site hosting on the market. This is just one illustration of how to achieve a long-standing public policy goal in the new environment created by the Internet.

Conclusion

Sitting to consider the application of First Amendment principles to the Internet, a federal court in Philadelphia found that the Internet is "the most participatory form of mass speech yet developed." However, the vast democratic potential of the Internet is only tentatively actualized in the United States, and far less so elsewhere. The extraordinary success of the Internet thus far, and the great potential that it holds out, goad us on toward making the Internet more widely available, and to doing the hard work of involving more people, and encouraging individuals and communities into the habit of political participation. For the great democratic potential of the Internet is not self-actualizing. Neither does the mere existence of the infrastructure guarantee that people will use it. But if they do, we may have a technology truly for the rest of culture.

The great student of American representative democracy Alexis de Tocqueville observed the essential connection between a decentralized communication, diversity of opinion, and democracy over one hundred and fifty years ago:

> [A]mong democratic nations the exercise of local powers cannot be entrusted to the principal members of the community as in aristocracies. Those powers must be either abolished or placed in the hands of very large numbers of men, who then in fact constitute an association permanently established by law for the purpose of administering the affairs of a certain extent of the territory; and they require a journal to bring them every day, in the midst of their own minor concerns, some intelligence of the state of their public weal (1835, p. 490).

The Internet presents us with an opportunity to support the highest goals of democracy. We ought to embrace the Internet and support its continued and growing us in political life.

Reference

Tocqueville, Alexis de, *Democracy in America*, Vol. 2, J.P. Meyer and Max Lerner, eds. (1835).

Technology and Capitalism

BY ROBERT HEILBRONER

IF there is one thing we have all learned from these pages, it is that the word *technology* has many facets. Yet, one aspect of this many, many-sided term deserves an examination it has not received. This is the relation between technology and capitalism. In this essay, I would like to sketch out some of the peculiar aspects of this relationship that warrant our attention.

Here I must begin by describing, in desperate brevity, what the word *capitalism* means. Capitalism is a "social formation," to borrow Marx's useful term, with three historically unique features: an all-important dependency on the successful accumulation of capital; a wide-ranging use of a market mechanism; and a unique bifurcation of power into two sectors, one public, one private. Together these institutional features serve both to guide the system in its daily workings as well as to maintain or change its long-term historical thrust. In so doing they radically alter the meaning and function of technology within capitalism compared with any other social order.

Before I turn to that central question, however, I must spend a few words discussing the properties of these all-important distinguishing attributes. Here it is usual to begin with the ubiquitous and pervasive presence of markets, both with regard to capitalism's day-to-day operations and its historic trajectory: indeed, one frequently hears the order called a "market system," perhaps to avoid having to pronounce its politically loaded real name. Yet, essential though the market is for capitalism, it is not here that we must look for its most remarkable relation to technology.

379

A second identifying element of a capitalist formation is its division of authority into two sectors: a public sector charged with the traditional duties and prerogatives of government, a private sector with the responsibility for producing the great bulk of total output. This bifurcation is also of great importance for capitalism, which in fact also likes to call itself the system of private enterprise; but important as the bifurcation may be both for political and economic reasons, it too is not the key aspect of the formation with regard to redefining the function of technology.

That leaves as the strategic point of examination the third unique feature, the pursuit of capital. Here I must begin by emphasizing the difference between capital and wealth. To put it as succinctly as possible, wealth is a thing, capital a process. Wealth consists of objects whose value lies in their symbolic embodiment of the power of their owner, and which are, therefore, never offered for sale except in dire emergencies. In sharp contrast, capital is embodied in commodities, such as coal or wool or labor power, that are unimaginable as wealth. More important, these commodities are thrown onto the market for more money than was spent in bringing them into being—not to pocket the receipts as wealth, but to buy or create more commodities which in turn will be used to repeat the process as long as it continues to generate more value. This is a dynamic circuit that has no counterpart in societies in the thrall of tradition or command. It is here that technology finds its historically unique role.

The nature of this new function must be obvious. In precapitalist social formations, improvements in the command over nature may have considerably improved the well-being of all, or the prestige of a few, but no one could claim that the very continuance of ancient or medieval or Renaissance life was under ever-present threat, or lived in ever-renewed hope, from the outcome of whatever technological change might be going on in its midst. By way of contrast, need I say that the outlook in all capitalist societies is in precisely such a state? Put differently, the invention of the spinning jenny and the steam engine, the steel mill and the gasoline motor, the generation of electricity and the design of the computer were never merely advances in the control over nature, but elements in an aspect of the enveloping social formation that had no counterpart in any precapitalist order. Prior to capitalism, improvements and deteriorations in material life were brought on by weather, military adventures, occasionally by political acts, but not by the dynamics of capital accumulation, which is to say, by self-generated changes in what would come to be called "economic life."

Technology thus becomes a sociopolitical force within capitalism, not merely a lever of material change. The reason, of course, is that technological change is the chief source of new areas of profitable accumulation. Here the market plays two roles, first in helping to facilitate the complex maneuvers by which the accumulation circuit works; later in bringing about the competition that will eat away at the profitability of these circuits. Capitalist economic history is thus written in bursts of accumulation largely brought on by technological change, followed by periods of slackening expansion as competition erodes profit margins.

This is not, of course, a steady or dependable process. Economic historian Joel Mokyr warns that technology is characterized by long periods of "stasis" (1990, p. 290f.). In addition, technological change can bring strain as well as stimulus: the introduction of new processes of production often renders obsolete the labor requirements of earlier processes: behind every ATM cash-dispensing machine one can make out the ghost of a bank teller. Nonetheless, speaking broadly, the steady search for and introduction of technology has been, and promises to be, the single most important source of capital accumulation, which is to say, of capitalist political security.

In addition, in a capitalist setting, technology also takes on a previously unknown responsibility for determining the social stability of its larger setting. The texture of daily life changes slowly, if at all, in hunting and gathering societies, and only gradually within the more dynamic societies of command. By way of familiar contrast, one of the most striking characteristics of capitalist life is the bewildering rate, and disconcerting depth, of social change—the conversion of *manu*-facture to *machino*-facture, with its immense repercussions on the activities called "labor"; the radical speeding-up of daily life as the telegraph and then the telephone annihilate distance; the lengthening of the day as electric light brings into being nightlife; the creation of a culture of senior citizens as pharmceuticals extend life spans. All this has endowed technology with sociological repercussions of immense magnitude and sweeping kind, although mostly unintended.

I hope I have made the case that the self-generated forces of capitalism endow technology with a sociopolitical importance far exceeding any it had previously enjoyed. Let me end by posing a much more daring question: What can be said with regard to the position that technology is likely to hold in the next social formation, assuming that capitalism will not last forever? The question becomes less of an excursion into science fiction when we recognize that we already stand within sight—indeed, within touch—of what will almost certainly be the dominant history-shaping force of the century ahead: global warming.

Global warming refers to the result of the steady accumulation of carbon dioxide in the atmosphere, where it serves as an invisible screen preventing the escape of heat generated by the normal reflection of the sun's rays, as well as that caused by combustion. I shall not stop here to cite the mounting scientific concern with respect to this relentless environmental change. Suffice it to note that the three warmest years since reliable record-keeping began in 1866 took place within the 1990s, with 1995 heading the list. There is growing fear that this unidirectional process could pose serious threats to human life within the coming century, and potentially disastrous ones for the century thereafter (Kennedy, 1993, p. 103; Brown et al., 1997, passim).

The thrust of these remarks is simple: technology as a means of extending humanity's control over nature has radically and irreversibly changed the relation between society and its erstwhile handmaiden. Capitalism has been the social formation that has brought this change about, but the change itself is today so deeply enmeshed in all societies, modern and would-be modern, that it will not disappear were capitalism itself to give way to some other, ecologically vigilant order. Tomorrow's—more accurately, the day-after-tomorrow's—technology will have to play the role, not only of designing, but of overseeing the operation of society's energy-related activities. As such, it seems likely to become the active locus of a widening human control over the forces of nature. I doubt, however, that it will provide the means to assure a necessary degree of control by society over its own behavior.

References

Brown, Lester, et al., *State of the World, 1997* (New York: Norton, 1997).

Kennedy, Paul, *Preparing for the 21st Century* (New York: Random House, 1993).

Mokyr, Joel, *The Lever of Riches* (New York: Oxford University Press, 1990).

Technology and Value

BY WILLIAM H. JANEWAY

I PROPOSE to identify three issues and address one. The first concerns the relationship between technological leadership, on the one side, and the frontier between the public and private sectors. The second calls into question the triumph of materialist rationality in our technological age. The third involves the relationship between the valuation of technology and the technology of valuation in an advanced capitalist regime.

(1) Two corporations share global leadership in today's age of information technology (IT): Microsoft and Intel. Neither company has benefited in any discernibly direct way from the financial or regulatory subvention of the state. I suggest that the rise of the joint "Wintel" enterprise to dominance represents a break with the history of technological innovation and economic development.

Most obviously, the success of Intel and Microsoft over the past decade owes nothing to direct government subsidy *versus* the case, most notably, of the railroads in the nineteenth century. In fact, Intel—the older and larger of the two—positively rejected Defense Department subsidies when it chose not to participate in the Very High Speed Integrated Circuit (VHSIC) program more than fifteen years ago, correctly recognizing that a focus on absolute performance would divert the company's best design talent from the challenge of optimizing price, performance, and compatibility with installed systems to achieve commercial success. Further, unlike the first generation of IT leaders in the twentieth century (such as IBM and Texas Instruments and the "national champion" IT companies of Europe and Japan), the growth of Microsoft and Intel was not materially supported by early and substantial demand from the public sector. With respect to less-direct modes of state support, the new IT leaders have neither lobbied for protection from foreign competition—such as nineteenth-century manufacturers—nor sought government-sanctioned predominance in the domestic market—such as the old AT&T.

On the contrary, Microsoft and Intel and the generation of IT innovators that have risen with the distribution of computing power since the early 1980's have committed themselves to living by and, often, dying by the sword of unconstrained competition. Thus, there is no little irony in the fact that the one immediately visible threat to their dominance—the Internet and the associated technologies brought to market most notably by Netscape—arise from an environment built and sponsored by the state. Only the resources and purposes of the national security state, arguably, could balance the cumulative power of positive returns to technological leadership and market share, driven by the explicitly "paranoid" genius of entrepreneurial leadership. As of this moment, it must be said, there is no reason to assume that the partners in the Wintel enterprise will not succeed in their radical attempts to reposition themselves for continued dominance on the new playing field open for them.

(2) The success of Microsoft and Intel expresses the reemergence of the old "virtues" of first-generation, liberal, free-market ideology. Each company is "open to the talents" with a vengeance. Rootless, in that recruits come from literally the entire globe, each is also ruthless in requiring success and dismissing failure. Unconstrained by notions of paternalist obligation to hazily characterized "stakeholders," Intel and Microsoft hire the best of those trained in the digital disciplines, offer them direct access to the wealth they collectively generate, and replace those who falter without qualms.

As economic institutions, these enterprises stand as triumphs of materialist rationality. How is it, then, that they have arisen in the one country of the developed world where an overwhelming majority of citizens affirm a faith in a personal creator? Is the emotional and spiritual toll of competing to win in the techno-logically driven arena such that it invites, even compels, commit-ment to transcendent, definitionally irrational beliefs? How else might that vast majority—whose conscious embrace of the values of the marketplace sanction the rise and rise of the Wintel enter-prise—cope with the certain knowledge that they themselves must fail the market's tests? Perhaps humankind can take just so much Schumpeterian creative destruction?

(3) Technology comes to market embedded in capital assets developed and distributed by enterprises whose securitized values, in turn, are themselves capital assets. The problematic process of valuing capital assets—tangible and intangible, illiq-uid and securitized—was at the core of Keynes's analysis of the dynamics of capitalism. In turn, the postwar triumph of neoclas-sical economics has depended on the credibility of its response to Keynes's critique.

In brief, the value of a capital asset is a function of the expected returns to ownership of it into the future compared with the cost of its production or purchase. Keynes argued that such expecta-tions were both fragile (subject to the famous "animal spirits" of entrepreneurs) and irrational (derived from the workings of the equally famous "casino" that is the stock market). The neoclassi-cal riposte to Keynes depends upon substituting the calculus of risk for emotional and inevitably inadequate efforts to overcome fundamental ignorance of future outcomes.

Technological innovation enters the argument on Keynes' side. History is replete with the inability to foresee what proved to be the commercially and financially meaningful applications of technological advance. One example may suffice for many: literally no one anticipated, when the science of the laser was being painfully reduced to reproducible practice, that, some twenty years later, the two dominant applications would lie in enabling optical communications and the automation of retail purchasing at the point-of-sale, respectively; some two hundred

start-up companies were born and died in the interim. Today, the
"Internet Boom" is merely another in a seemingly endless
sequence of stock market bubbles as investors speculate on each
other's fear of missing the "next great thing." A rapid pace of
technological advance, as it renders obsolete sunk investments
and opens the door for an Intel and a Microsoft, can confidently
be expected to increase the spread of expected returns and
reduce the time horizon over which the rational agent is prepared to express any expectation at all.

Yet, if the process of valuing technology is undeniably problematic, perhaps technology can be turned to transform the
methodology of valuation itself. Information technology has done
more than bring close the frontier of investment ignorance. It has
itself enable a vast increase in the scale and scope of the markets
in which financial assets are traded. For at least a hundred years,
since Léon Walras, the dream of the neoclassically minded economist has been to complete the system of markets necessary to
define a general equilibrium, such that at all moments all prices
are set at market-clearing levels in all markets. Today's IT offers
the promise of accomplishing the dream: this is what the endless
proliferation of financial derivatives, defined and traded by computer programs, seems to represent.

In fact, the growth of derivatives is not a function of the logical
needs of an economic theory. And a good thing, too: solving the
theoretical challenge is a logical—not a technological—impossi-
bility as it requires, of course, not only an infinite set of futures
markets, but an (infinitely larger) infinite set of *contingent* futures
markets. Rather, derivatives represent the pragmatic desire of
market participants to cope with ignorance by laying off risk
through the creation of instruments which capture specific "bun-
dles of risk." Alas! The first dramatic consequence of the
systematic, computerized effort of capital market players to lay off
risk was the rise of "portfolio insurance" in the 1980s and its
denouement in the crash of 1987. By attempting to use the derivative markets in stock market indices to lay off the risk of owning

portfolios of equities, market participants created an implosion of value. Seldom has the Law of Unintended Consequences asserted itself so decisively.

The crash of 1987 and a host of microcalamities in the derivative markets serve to confirm another and related assertion of Keynes, one that was rooted in his own experience of markets as an active player. Sooner or later, the rationally calculated "value" of any asset is subject to the test of liquidity: at what price can it be exchanged for ready money? The "technological fix" of portfolio insurance failed disastrously in October 1987, as computer programs responded instantaneously to the actions of computer programs, with no time lag allowed for human judgment to assert itself. The collapse of the time constant, and the consequent elimination of market liquidity, has been effectively dealt with by the most simple and nontechnological device imaginable: the imposition of "circuit breakers" as trading halts whenever specified threshold levels of decline in stock market prices are reached. The time for reflection imposed from without the market, simply put, allows liquidity to catch up with trading.

The net of all this is to suggest that "technology" is not an escape hatch from the conditions of market existence. On the contrary, technology in its economic aspects is embedded in the problematic dynamics of the market economy, where greed and fear contend in the face of the unforeseeable.

Liberal Maps for Technology's Powers: Six Questions

BY IRA KATZNELSON

I WANT to inquire about the relationship between two not entirely autonomous developments: the radical transformations convened by high technology, especially to the means and terms of communications, and the victory of liberal over totalitarian modes of political thought and governance. Arguably, the fall of Bolshevism, totalitarianism's last large-scale redoubt, was caused by its illiberalism; that is, by its inability to provide the free space required to grapple with the challenges of producing and living with our era's new technologies. Communism failed the tests of economic restructuring, functional interconnectedness, cross-border flows of ideas, the pressurized acceleration of time, the contingency of space, and the plasticity of identity. Its dispirited elites understood their incapacities in the face of these revolutionary transformations. Demoralized, they abdicated and liberalism triumphed, at least provisionally.

There is something of an elective affinity between liberal political thought and the fragmentation, discontinuity, eclecticism, and ephemerality characteristic of an epoch defined by flexible technologies. Liberalism, after all, has valued and sought to sustain open societies and the free exchange of information and views. Notwithstanding, the political and ethical tools liberalism offers to survey and guide our engagement with the technological revolution provoke difficult, intertwined questions, including the following six, regarding the character, capacities, and choices of liberal theory.

1. Inclusion

How much of humankind will be placed inside the new technology's, hence liberty's, sphere in light of current radical shifts? For John Locke, issues of inclusion and exclusion hinged on the capacity of the human person to be a rational, thinking agent. John Stuart Mill, by contrast, authorized criteria for inclusion based on social and cultural development, thus on what he saw as a contrast between backwardness and civilization. The new technologies of communication are sharply eroding many markers of particularity that have defined many of this century's divisions: between nations and language groups, for example; but they also are exacerbating axes of inequality coinciding with hierarchies of wealth, geography, and positionality in networks of exchange. As a result, Locke's focus on rationality and Mill's on unenlightened peoples and places now combine to produce new mass underclasses of excluded people. In a context where what it means to be a rational, autonomous agent has become more demanding and where access to technologically advanced civilization has become more skewed by fresh accumulations of advantage, how can liberalism sustain its twentieth-century impulse favoring incorporation and democratic citizenship? Or are we doomed, by technological fiat, to return to a far less inclusive, more stratified liberal vision?

2. Agency

Liberalism often is portrayed as the partner of science and technological advance, just as the scientific revolution in early modern Europe is characterized as providing the handmaiden of the Enlightenment. There is, however, quite a deep tension between the mechanization and mathematization of nature conceived in homogeneous and abstract space by the physical sciences and the central idea of Enlightenment reason by individuals who become authors of their destiny. Likewise, the modern biological sciences, certainly since Darwin, have inclined toward agentlessness. It is something of a truism these days that the new technology is shaping a world that effectively appears outside human control, in part because it spans familiar boundaries of place and political control; in part because it multiplies and intensifies experiences, collapses time and historical perspective, and makes change more unpredictable. In such settings, human agency and consent, especially in the political realm, seem increasingly elusive and remote; but without such agency and consent, liberal theory is made a mockery.

3. Sovereignty

Because the new technology has transformed the velocity and location of economic processes, especially the scale and celerity of financial markets and the scope of networks conveying information, the conventional boundedness of states claiming sovereignty over people inside finite borders is called into question. Liberal theory has been premised, however, on just this world of states created in the early modern West in the aftermath of feudalism. To the extent the ensemble of institutions possessed by sovereign states governed by liberal visions of public authority, governance, and representation is incapable of effectively managing this new state of affairs, they face tests of potential demoralization comparable to those already failed by the globe's most illiberal regimes.

4. State Power

As Steve Shapin reminds us in his recent provocative book on the scientific revolution (1996), modern scientific knowledge from the start was thought to be pertinent to political problems of power and order. Francis Bacon, for example, who counseled Queen Elizabeth I and King James I and served as lord chancellor, thought it urgent to find a balance between free inquiry and control of knowledge; between the reform of learning and the creation of conditions under which reformed learning would enhance state power and be disciplined by its imperatives. In his

New Atlantis (1627), Bacon described Solomon's House, an engineering and research institute serving the interests of the state in the imaginary world of Bensalem, whose work, Shapin notes, was two-fold: "the extension of natural philosophy (the knowledge of causes) and the extension of power ('The enlarging of the bounds of human empire')" (Shapin, 1996, p. 130). Bacon understood that knowledge could be an instrument of centralized power, that states abdicated their capacity to monitor the creation of knowledge at great risk, and, as he famously put it, that "human knowledge and human power meet in one." Legitimated this way, he argued that science and technology have just claims on the public purse. Eyeing these issues from the perspective of liberal thought and norms, we need to inquire anew about these issues to identify a repertoire of relationships joining our new technologies with the capacities of today's states in normatively desirable ways.

5. Ethical Narration

Liberalism provides more than a set of political guidelines. It also convenes an ethical imaginary; really a multiplicity of such imaginaries that presuppose a significant degree of common experience, cognition, and networks of obligation and exchange and that mediate between the impersonal and the local. Ethics also presuppose an immediacy of time and space in which these elements converge (see Tester, 1992, p. 207). But just the boundaries required for ethical thought and action are being eroded by the new technologies. In consequence, aesthetics and emotiveness threaten to replace ethics, spectacle to replace deliberation, and a hyperrelativism to replace demanding standards in human affairs and in the quest for knowledge, with potentially fateful results for political liberalism.

6. Public Opinion

Taken together, the challenges of inclusion, agency, sovereignty, state power, and ethical narration endanger political liberalism's central idea that government be based on informed consent by publics guided by opinion, which, deployed through such instruments as political parties, pressure groups, elections, and legislative representation, could shape regimes where opinion, however complex and conflictual, is influential if not controlling. The political scientist, V.O. Key's important contribution to the empirical and normative study of these issues defined public opinion "to mean those opinions held by private persons which governments find it prudent to heed" (Key, 1961, p. 14). Focusing on the formation of opinion by the family, the educational system, and the media, as well as by informed political elites, he cautioned about the decay of liberal democracy when these processes come to be misshaped and contused. Clearly, the new technologies augment the ready flow and scale of information and the available range of opinions. But they also enhance prospects for marketing and manipulation, substituting the orchestration of aesthetics for reason. From a liberal perspective, no question is more urgent than that which asks how the new technology might, in spite of these dangers, contribute toward the creation of thoughtful, informed opinion in the sense of opinion that cannot be ignored.

References

Key, V.O., Jr., *Public Opinion and American Democracy* (New York: Alfred A. Knopf, 1961).

Shapin, Steven, *The Scientific Revolution* (Chicago: University of Chicago Press, 1996).

Tester, Keith, *The Two Sovereigns: Social Contraditions of European Modernity* (London: Routledge, 1992).

Notes on Contributors

Robert McC. Adams
is Adjunct Professor of Anthropology at the University of California, San Diego and Secretary Emeritus of the Smithsonian Institution. His most recently published work is *Pathos of Fire: An Anthopologist's Inquiry into Western Technology* (1996). His previous publications have dealt primarily with the long-term urban and agricultural development of the Near East.

Jerry Berman and Daniel J. Weitzner
are Executive Director and Deputy Director, respectively, of the Center for Democracy and Technology. Their most recent publication is *Abundance and User Control: Renewing the Democratic Heart of the First Amendment in the ASC of Interactive Media*. They are also authors of the two successful Supreme Court challenges to the Communications Decency Act: *Reno vs. ACLU*. This paper is based, in part, on a presentation delivered by Mr. Weitzner at the Academy for the Third Millenium's Conference on Internet and Politics. The web page for the Center for Democracy and Technology is http://www.cdt.org.

Elizabeth L. Eisenstein
is Professor Emerita in the History Department at the University of Michigan, Ann Arbor. Her publications include *The Printing Revolution in Early Modern Europe* (1993), *Grub Street Abroad* (1992), and *The Printing Press as an Agent of Change* (1979). Dr. Eisenstein is currently working on *Divine Art/Infernal Machine: Western Views of Printing Surveyed.*

Peter Galison
is Mallinckrodt Professor of History, of Science, and of Physics at Harvard University. His publications include *Image and Logic: A Material Culture of Microphysics* (1997) and *The Disunity of Science*, coedited with David Stump. He is working on a history of objectivity with Lorraine Daston and has begun a study of postwar quantum field theory. The article in this issue of *Social Research* draws heavily on Dr. Galison's *Image and Logic*, just out from Chicago University Press.

Paul Gewirtz
is Potter Stewart Professor of Constitutional Law at Yale Law School. He is the author of *Law's Stories: Narrative and Rhetoric in the Law*, with Peter Brooks (1996) and "The Triumph and Transformation of Antidiscrimination Law."

Robert L. Herbert
is Andrew W. Mellon Professor of Humanities Emeritus at Mount Holyoke College. Dr. Herbert is the author of "'Architecture' in Leger's Essays 1913–1933," in *Architecture and Cubism* (1997) and *Monet on the Normandy Coast* (1994). He is currently working on *Renoir's Doctrine of Irregularity, The Artist's Writings on the Decorative Arts* (1998).

Robert Heilbroner
is Norman Thomas Professor of Economics Emeritus at The New School for Social Research. Dr. Heilbroner is the author of *Teachings from the Worldly Philosophy* (1996).

John Hollander
is Sterling Professor of English at Yale University. His most recently published work is *The Gazer's Spirit: Poems Speaking to Silent Works of Art* (1995).

Nicholas Humphrey
is Professor of Psychology at The New School for Social Research. His publications include *Leaps of Faith* (1996) and *A History of the Mind* (1992).

William H. Janeway
is the Managing Director at E. M. Warburg, Pincus & Co., LLC. His most recently published work is "The 1931 Sterling Crisis and the Independence of the Bank of England" in the *Journal of Post Kenesian Economics.*

George Kateb
is Professor of Politics at Princeton University. Dr. Kateb is the author of *Emerson and Self-Reliance* (1995), *The Inner Ocean: Individualism and Democratic Culture* (1992), and *Hannah Arendt: Politics, Conscience, Evil* (1984).

Ira Katznelson
is Ruggles Professor of Political Science and History, Columbia University. His most recent publication is *Liberalism's Crooked Circle: Letters to Adam Michnik* (1996).

Joshua Lederberg
is Sackler Foundation Scholar and President Emeritus at the Rockefeller University, New York.

Leo Marx
is Kenan Professor of American Cultural History, Program in Science, Technology, and Society at Massachusetts Institute of Technology. He is the author of *Progress: Fact or Fiction?* (1996) and "Does Improved Technology Mean Progress" published in *Technology Review* (1987).

Marvin Minsky
is Toshiba Professor of Media Arts and Sciences at Massachusetts Institute of Technology. His most recently published work is *The Turing Option* (a novel with H. Harrison, 1992). He is also the author of *The Society of the Mind* (1987), *Robotics* (1986), *Artificial Intelligence* (1972), Perceptrons (1969, enlarged edition 1988), and *Semantic Information Processing* (1968).

David E. Nye
is Professor of American Studies at Odense University in Denmark. His most recently published works are *Consuming Power: A Social History of American Energies* (1997) and *Narratives and Spaces: Technology and the Construction of American Culture* (1997). He is also the author of *American Technological Sublime* (1994), *Electrifying America* (1990, Dexter Prize, Abel Wolman Award), and *Image Worlds* (1985). Dr. Nye is currently working on *Narrating Power.*

Arno Penzias
is Vice President and Chief Scientist at Bell Labs, Lucent Technologies. He is the author of *Digital Harmony* (1995) and *Ideas and Information* (1989) and is currently working on "The Next Fifty Years," *Bell Laboratories Technical Journal.*

Alan Ryan
is Warden at New College, Oxford University. His most recent book is *John Dewey* (1995).

Alan Trachtenberg
is Neil Gray, Jr. Professor of English and American Studies at Yale University. His most recently published work is *Reading American Photographs* (1989). Dr. Trachtenberg is also the author of *The Incorporation of America* (1982) and *Brooklyn Bridge: Fact and Symbol* (1965).

Sherry Turkle
is Professor of Sociology of Science at Massachusetts Institute of Technology. She is the author of *Life on the Screen: Identity in the Age of the Internet* (1995), *Psychoanalytic Politics: Jacques Lacan and Freud's French Revolution* (1991), and *The Second Self: Computers and the Human Spirit* (1984).

Rosalind Williams
is Dean of Students and Undergraduate Education and the Metcalfe Professor of Writing at Massachusetts Institute of Technology. She is the author of *Notes on the Underground: An Essay on Technology, Society and the Imagination* (1990) and *Dream Worlds: Mars Consumption in Late 19th Century France* (1982). She is currently working on *The Roots/Routes of Modern Life: Studies in Geography and Imagination.*

Langdon Winner
is Professor of Political Science and is Director of Graduate Studies in the Department of Science and Technology Studies at Rensselaer Polytechnic Institute. Dr. Winner is the author of *The Whale and the Reactor: A Search for Limits in and Age of High Technology* (1986), *Autonomous Technology: Technics-out-of-Control as a Theme in Political Thought* (1977). He is currently working on *Political Artifacts: Design and the Quality of Public Life.*